Learning from Failures

Learning from Failures

Learning from Failures
Decision Analysis of Major Disasters

Ashraf Labib

AMSTERDAM • BOSTON • HEIDELBERG • LONDON
NEW YORK • OXFORD • PARIS • SAN DIEGO
SAN FRANCISCO • SINGAPORE • SYDNEY • TOKYO

Butterworth-Heinemann is an imprint of Elsevier

Butterworth-Heinemann is an imprint of Elsevier
The Boulevard, Langford Lane, Kidlington, Oxford OX5 1GB, UK
225 Wyman Street, Waltham, MA 02451, USA

First edition 2014

Notice
Knowledge and best practice in this field are constantly changing. As new research and experience broaden our
understanding, changes in research methods, professional practices, or medical treatment may become necessary.

Practitioners and researchers must always rely on their own experience and knowledge in evaluating and
using any information, methods, compounds, or experiments described herein. In using such information or
methods they should be mindful of their own safety and the safety of others, including parties for whom they
have a professional responsibility.

To the fullest extent of the law, neither the Publisher nor the authors, contributors, or editors, assume any
liability for any injury and/or damage to persons or property as a matter of products liability, negligence
or otherwise, or from any use or operation of any methods, products, instructions, or ideas contained in the
material herein.

British Library Cataloguing-in-Publication Data
A catalogue record for this book is available from the British Library

Library of Congress Cataloging-in-Publication Data
A catalog record for this book is availabe from the Library of Congress

ISBN: 978-0-12-416727-8

For information on all Butterworth-Heinemann publications
visit our web site at books.elsevier.com

14 15 16 17 18   10 9 8 7 6 5 4 3 2 1

Contents

Acknowledgments

I would like to thank the many people who have motivated, inspired and provided me with valuable information and ideas that have contributed to this book. My mother, Prof (Emiritus) Aida Nossier-Labib, for her encouragement, Mr. John Harris for editing the whole manuscript, Fiona Geraghty, and Cari Owen from Elsevier for their professional support. Also, special thanks to Ms. Pauline Wilkinson – Project Manager from Elsevier for making the final proofs.

I am also grateful to my colleagues and students who have contributed to the different chapters of the book. For Chapter 4, I am grateful to John Unwin, Paul McGibney, and Mazen Al-Shanfari, from University of Glasgow Caledonian, who helped to collect the data for the Bhopal case study. I am also grateful to Rebecca Harris for the proof reading and editing. I am also grateful to John Harris for his valuable feedback and comments. I am grateful to my co-authors Ramesh Champaneri and Alessio Ishizaka for the two related papers co-authored ion this subject.

For Chapter 5, I am grateful to my students at both University of Manchester (MSc Reliability Engineering) and University of Portsmouth (MBA). Specifically, those who have worked on the Deepwater Horizon Case study: Lim Chung Peng, Oluwafemi Akanbi, Richard Odoch, Amiel Rodriguez Adrian Nembhard, Saeed Alzahrani, Igwe Efe, Isip Ime, Shamsul Amri, Joledo Akorede, and Luwei Kenisuomo.

For Chapter 6, I am grateful to my students at both University of Manchester (MSc Reliability Engineering) and University of Portsmouth (MBA). Specifically, those who have worked on the BP Oil Refinery Disaster Case study: Mod Bin Mazid, Ken Hurley, Shehzad Khan, Akif Najafov, and Hernan Quiroz.

For Chapter 9, I am grateful to my students at both University of Manchester (MSc Reliability Engineering) and University of Portsmouth (MBA). Specifically I am grateful to the two groups involved in the Fukushima project and in particular, the leaders of the groups; Precious Katete and Michael Booth.

I am also grateful to Professor Chris Johnston and Dr. El Sobky for the valuable information and criticism they provided me with. My appreciation also goes to Dr. Martin Read for co-authoring related papers with me.

My great appreciation and thanks go to my family; my wife and my son for their patience and encouragement, without them this book would not have materialised.

Part 1

Background of Analytical Methods Used in Investigation of Disasters

1

Background of Analytical Methods Used
in Investigation of Disasters

Introduction to the Concept of Learning from Failures

1.1 **INTRODUCTION**

There is evidence that lessons gained from major disasters have not really been learned by the very same organizations involved in those disasters: the multiple disasters, for example, that occurred in organizations such as NASA (the Challenger and Columbia accidents) and BP (Texas City refinery and Deepwater Horizon accidents). So why do organizations and institutions fail to learn? When accidents happen, what are the factors that can drive the unlearning process? And, how can organizations learn and change their policies, routines, and procedures through feedback? In this book, the concepts of learning and unlearning from failures are investigated and a new theory is developed in order to address these questions and to provide a mechanism for feedback. It has been reported that organizations learn more effectively from failures than from successes (Madsen and Desai, 2010) that failures contain valuable information, but that organizations vary in their ability to learn from them (Desai, 2010). It is also argued that organizations vicariously learn from the failures and near-failures of others (Kim and Miner, 2007; Madsen, 2009). However, it can also be argued that lessons gained from major failures have often not really been learned by the very same organizations involved in them. This been exemplified by recent incidents within major organizations such as BP, NASA, and Toyota.

The first case concerns BP. In March 2005, a series of explosions and fires occurred at its Texas City refinery killing 15 people and injured 170 (Vaidogas and Juocevičius, 2008). An analysis of BP's recent history (Khan and Amyotte, 2007) showed that the March 2005 disaster was not an isolated incident and concluded that BP led the US refining industry in its incidence of fatalities over the previous decade—and in April an

Learning from Failures. DOI: http://dx.doi.org/10.1016/B978-0-12-416727-8.00001-1

explosion destroyed its Deepwater Horizon drilling rig, killing 11 workers and initiating a major oil spill.

The second case concerns NASA which experienced the Challenger launch disaster in 1986 followed by the Columbia disaster in 2003. Both failures have been analyzed (Vaughan, 1996, 2005) and attention drawn to a consistent and institutionalized practice of underestimating failures as early warning signals and having too much belief in the track record of past successful launches. It was also noted that NASA concluded that both accidents were attributed to "failures of foresight" (Smith, 2003).

The third case concerns Toyota. In January 2010, a quality problem affected Toyota which led to a global recall, of more than 8.5 million vehicles, to deal with various problems, including sticking gas pedals, braking software errors, and defective floor mats. And again, on July 5, 2010, Toyota began recalling more than 90,000 luxury Lexus and Crown vehicles in Japan as part of a global recall regarding defective engines (Kageyama, 2010).

The author has coauthored a paper entitled *"Not Just Rearranging the Deckchairs on the Titanic: Learning from Failures through Risk and Reliability Analysis"* published in *The Journal of Safety Science* (Labib and Read, 2013), which is the first part of the title implying that learning should not lead to doing something pointless or insignificant that will soon be overtaken by events, or that contributes nothing to the solution of the current problem.

I have also written and coauthored related papers which have followed the same theme as in this present book in proposing an analytical tool or a hybrid of such tools in an integrated approach and then demonstrating the value of such an approach through a case study of a disaster or of multiple disasters. This included one (Davidson and Labib, 2003) in which we analyzed the 2000 Concorde accident using a multiple criterion prioritization approach called the analytical hierarchy process (AHP). There, we also proposed a systematic methodology for the implementation of design improvements based on experience of past failures and this has been conducted and applied in the case of the Concorde after the 2000 accident.

I have also been involved in presenting reliability engineering techniques such as failure mode effect analysis (FMEA), fault tree analysis (FTA), and reliability block diagrams (RBD) which have been used to analyze the Bhopal disaster (Labib and Champaneri, 2012) and show how such techniques can help in building a mental model of the causal effects of the disaster. The Bhopal study was also used to develop a new logic gate in the fault tree proposed for analyzing such disasters and to show the

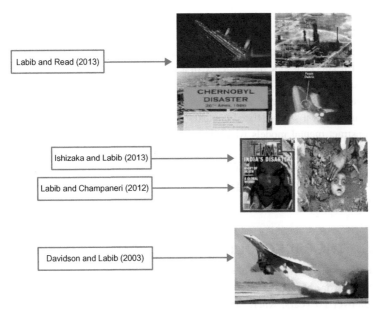

■ **Figure 1.1** Learning from failures using systematic approaches. Papers published by the author related to this book.

benefits of using hybrid techniques of multiple criteria and fault analysis to evaluate and prevent disasters (Ishizaka and Labib, 2013).

Four case studies of disasters from different industries have also been studied: The Titanic, Chernobyl, BP Texas city refinery, and NASA's Columbia Shuttle (Labib and Read, 2013). The root causes of the disasters were analyzed, firstly via a description of the sequence of events that led to the disaster and then via an investigation of the technical and the logical causes of the failure as well as of its consequences. A set of 10 generic lessons was then extracted and recommendations made for preventing future system failures (Figure 1.1).

So, why do failures happen? Are organizations really learning from failures? And in the context of a failure, how is learning realized? These are the questions which I shall try to address.

1.2 **WHY LEARNING FROM FAILURES?**

Disasters destroy not only lives but also reputations, resources, legitimacy, and trust (Weick, 2003). It has been argued that system accident literacy and safety competence should be part of the intellectual toolkit of all engineering students (Saleh and Pendley, 2012). Safety is a fascinating and significant subject that lies across different disciplines, yet the field still suffers from fragmented literature on accident causation and system safety (Saleh et al., 2010). This is despite the existence of a set of well-developed theories in the field (Pidgeon, 1998). In learning about how to learn from failures, it is both necessary and timely to appreciate the different facets of this field. Learning, in a safety related context, is about feedback that can help us to construct mental models which in turn can lead to taking better decisions (Saleh and Pendley, 2012). A framework of the concept of learning from failures was proposed by Labib and Read (2013) which relies on three principles: (1) feedback to design, (2) use of advanced techniques for analysis, and (3) extraction of interdisciplinary generic lessons.

1.3 **THE BACKGROUND OF THIS BOOK**

The book is based on material that the author has used in teaching courses on reliability and maintenance engineering at various universities for over 20 years, where the proposed models and case studies covered were presented and also draws on material from the research carried out by the author and his coauthors. I have taught Masters Programs in the field of Reliability and Maintenance at Manchester and Glasgow Caledonian Universities. The author is regularly invited to the Middle East, Europe, Canada, and the United States to speak about the recent developments of such models.

This work originated in the author's involvement in teaching a course in Design for Maintainability, which is part of the University of Manchester's Master Program in Reliability and Maintenance Engineering. The students, who were mainly from industry, were asked to use reliability engineering and maintenance management techniques to analyze the root causes of major machine failures, or disasters, and extract generic lessons that can feedback to design. Here, design was loosely defined to mean not just the design of particular equipment but also the design of new procedures and any recommendations that can help to prevent problems from reoccurring. This subject expanded and started to be included in industrial workshops and training as well as in MBA Master Classes that were conducted at the University of Portsmouth, UK, in the area of organizational learning from failures.

1.4 **WHO SHOULD USE THIS BOOK**

The book is applicable for universities at postgraduate level and to various practitioners such as safety, risk, quality, and reliability managers and officers. It is aimed at both postgraduate (PG) and senior undergraduate (U/G) students in engineering and management programs and introduces the student to the framework and interdisciplinary approach at a level appropriate to the later stages of an undergraduate program. Also, the book links theory with practice and exposes the student to the evolutionary trends in risk, safety, and reliability analyses. As a result, students after graduation should have sufficient skills to work in industry such as reliability, maintenance, safety, and quality professionals. This background will also motivate those pursuing further studies and research to consider these fields as an interesting and challenging area for their future career.

1.5 INTRODUCTION TO THE CONCEPT OF LEARNING FROM FAILURES

Questions within the process study field concern the evolutionary process, which determine how and why things emerge and develop. This is different from questions of variances, which focus on covariations among dependent and independent variables. An example of process studies at the organization level is the Balogun and Johnson (2004) research into how middle managers make sense of change as it evolves. Another example is the work of van de Ven and Poole (1995) in which the authors focus on the temporal order and sequence in which selected managerial or organizational phenomena develop and change, which contributes to the process theories of organization and management. Tsoukas and Chia (2002) suggest that processes of change are continuous and inherent to organizations. This is particular helpful when investigating how failures develop and the learning that can then occur. In this book, there is an attempt to address a strand of process studies of organizational innovation and change that examines how individuals and organizations learn from failures, and how disasters unfold over time.

Within the field of organizational learning, there is a view that incremental learning tends to be control oriented and which therefore seeks to maintain predictable operations, to minimize variation, and to avoid surprises, whereas radical learning tends to seek to increase variation in order to explore opportunities and challenge the status quo (Carroll et al., 2002). Failures are enablers for change in the status quo. They challenge the status quo and old assumptions. They force decision makers to reflect on what went wrong, why, and how (Morris and Moore, 2000).

The traditional school of thought tends to emphasize learning from successes (Sitkin, 1992; McGrath, 1999; Kim and Miner, 2000), whereas more research has begun to explore whether organizations can also learn from failures of other organizations, or from failures that occur within the same organization (Carroll et al., 2002; Haunschild and Sullivan, 2002; Denrell, 2003; Haunschild and Rhee, 2004; Chuang et al., 2007; Kim and Miner, 2007; Desai, 2008; Madsen and Desai, 2010). Experience with failure has proved to be more likely than experience with success to produce conditions for experiential learning, the driver to challenge existing knowledge, and the ability to extract meaningful knowledge from experience (Haunschild and Rhee, 2004; Madsen and Desai, 2010).

Until now, mainstream research on organizational learning has treated the impact of rare events as of only marginal interest, such events being set aside as statistical outliers (Lampel et al., 2009). However, a special issue of Organization Science in 2009 was dedicated to this topic. In this issue, Lampel et al. proposed a taxonomy of learning that was based on two categories: the impact of rare events on the organization and their relevance to that organization. Based on this, they identified four classes of events: transformative, reinterpretative, focusing, and transitory. They then mapped the works of Beck and Plowman (2009), Christianson et al. (2009), Madsen (2009), Rerup (2009), Starbuck (2009), and Zollo (2009) in to those classes. This book is about disasters that have had high impact and hence they may be regarded as either transforming or focusing, depending on their degree of relevance to a specific organization. However, there is also a provision of generic lessons that are applicable to a wide range of industries.

It is generally accepted that learning from failures is a difficult process to assess. Few authors have attempted to define it. Organizational learning has been defined by Madsen and Desai (2010) as any modification of an organization's knowledge occurring as a result of its experience. But again it is acknowledged that a change in organizational knowledge is itself difficult to observe. Subsequently, there has been a trend in research which argues that changes in observable organizational performance reflect changes in organizational knowledge (Argote, 1999; Baum and Dahlin, 2007).

Another line of research attempts to explore ways of learning. For example, Carroll et al. (2002) proposed a four-stage model of organizational learning reflecting different approaches to control and learning. Also, Chuang et al. (2007) developed a model of learning, from the failures of health-care organizations, that highlights factors that facilitate such learning.

Learning from disasters in the context of socio-technical system failures was studied in the pioneering work of Turner (1978) on man-made disasters, by Toft and Rynolds (1997) in their identification of a topology of isomorphism, by Shrivastava et al. (1988) in their analysis of symptoms of industrial crises and their proposal of a model of socio-technical crisis causation based on either "HOT" event initiators of failures (human, organizational, and technological) or "RIP" events accelerators of failures (regulatory, infrastructural, and political). This was then extended by the work of Smallman (1996) in challenging traditional risk management models, Pidgeon and O'Leary (2000) in investigating the interaction between technology and organizational failings, and Stead and Smallman (1999) in comparing business to industrial failures, where the issue of interaction between technology and organizations was further analyzed and applied to the banking sector.

Learning from failures has been researched as an outcome rather than as a process itself. It has been argued, by Madsen and Desai (2010), that learning from failures is related to improved performance. It has also been argued by Barach and Small (2000) that learning from near-misses helps in redesigning improving processes. Learning has been defined by Haunschild and Rhee (2004) to be, in the context of car recalls, a reduction in subsequent voluntary recalls. Stead and Smallman (1999) proposed the process of a crisis as an event cycle consisting of a set of *preconditions*, a *trigger* event, the *crisis* event itself, and a period of *recovery*, which then leads to the *learning* phase. They attempted to ground this "crisis cycle" theory on the works of Turner and Pidgeon (1997), Shrivastava et al. (1988), Pearson and Mitroff (1993), and Pearson and Clair (1998).

Case studies of accidents have been presented by Kletz (2001) in an effort to learn from accidents by carrying out a thorough analysis of the causes of the failures. Therefore, the majority of the related literature, apart from that mentioned above, tends to focus on learning from failure as an outcome rather than on why disasters occur. Also, there is a lack of research work on how the learning process can emerge or on how to use models that can facilitate the process of learning from failures and extracting generic lessons. There is an argument that using a model to analyze and investigate an accident may distract people who are carrying out the investigation, as they attempt to fit the accident to the model, which may limit free thinking (Kletz, 2001). I tend to agree with this and have therefore argued here that the use of a hybrid of models overcomes this issue as such an approach will lead to limiting the assumptions inherit in such models an argument which will be enlarged upon in later chapters.

The above discussion shows that the process of learning from failures is multilayered and its understanding involves a variety of related theories.

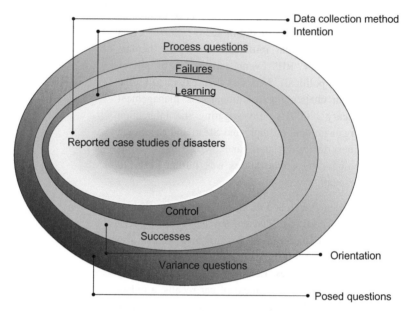

■ **Figure 1.2** Research focus. *From Saunders et al. (2003).*

Figure 1.2, based on the "research onion" model of Saunders et al. (2003), shows how the scope of this book fits with respect to those theories. To summarize, the proposed approach is based on process oriented questions rather than variance questions, focuses on radical learning rather than control learning, and is based on failures rather than on successes.

The next section introduces a taxonomy of the theories relating to learning from failures and reviews the importance of having a balanced approach toward those theories.

1.6 **TAXONOMY OF THEORIES**

In the following subsections, an attempt will be made to construct a taxonomy of the various theories around learning from failures in a pairwise style, i.e., involving the presentation of two seemingly opposed, or paradoxical, theories with respect to a certain topic that relates to safety analysis. By using such an approach, it is hoped that the reader will develop a balanced understanding of the theories, one which will enhance his skill in both formulating and analyzing the factors that have led to disasters and the issues related to lessons learned to prevent reoccurrence of any similar disasters.

1.6.1 **Learning from Case Studies Versus the Narrative Fallacy Concept**

Learning from reported case studies of carefully chosen major disasters is based on the arguments of Greening and Gray (1994) and Desai (2010), in that catastrophic failures are well observed and this visibility tends to encourage organizations as well as policy makers, to learn from those incidents. Although this methodology is limited by its reliance on secondary research, i.e., investigation carried out by others, it provides a fascinating body of research in that, as explained by Wood et al. (1994): "*We most fully integrate that which is told as a tale.*" The analysis of these case studies is important for demonstrating how subjectivity that relies on the opinions of experts can be turned into a modeling approach that can ensure repeatability and consistency of results. However, one should rely, as much as possible, on facts and avoid being biased by the framing of the narrative of the case study.

The concept of the "narrative fallacy" was introduced by Taleb (2010) to describe how flawed stories of the past shape our views of the things around us. Narrative fallacies arise from our continuous attempt to make sense of the world. Taleb suggests that we humans constantly fool ourselves by constructing biased accounts of the past and believing they are true. According to Kahneman (2012), this misleading assumption that the past has been fully understood can feed into a further illusion that everything is simple, predictable, and coherent, and that the future can be predicted and controlled. Kahneman argues that this tendency to provide false reassurance has been observed in many business books designed to satisfy this particular need. This is in line with the *reification fallacy* introduced by Gould (1981), which was described by Hollnagel et al. (2006) as "*the tendency to convert a complex process or abstract concept into a single entity or thing in itself.*" Therefore, one could argue that, when presenting a disaster in the form of a case study, one needs to be aware of the possibility of narrative fallacy and when possible assign a confidence level to the arguments presented, especially if they are not based on factual evidence. One can also look at the same problem from different perspectives using more than one mental model or incorporating multiple techniques for the analysis.

1.6.2 **Learning from Hindsight Rewards Versus Risk Aversion**

When dealing with attitudes toward the perception of risky decisions and safety measures, there are two extreme approaches that need to be avoided. One, as proposed by Sunstein (2005), is intense risk aversion, where the precautionary principle is costly and, when interpreted strictly, it can be paralyzing. Sunstein mentions a list of innovations that would not have passed the test, including *"airplanes, antibiotics, open-heart surgery . . . and X-rays."* The other is outlined by Kahneman (2012) and is related to hindsight and its outcome bias that may reward irresponsible risk seekers. For example, leaders who have been lucky are never punished for having taken too much risk. On the contrary, they are believed to have had the flair and foresight to anticipate success, while those sensible people who doubted them are seen in hindsight as mediocre, timid. and weak. A few lucky gambles can crown a reckless leader with a halo of boldness and an incentive to take more risky decisions. The same view was echoed by Starbuck and Milliken (1988) who argued that retrospective analyses always oversimplify the connections between behaviors and outcomes, and make the actual outcomes appear highly inevitable and highly predictable.

Reason (1997) proposed what came to be known as the Swiss cheese model, which attempts to describe the occurrence of an accident as a ray of light that manages to penetrate the holes of the many existing parallel safety related barriers (Swiss cheese slices). Sunstein's paralyzing risk aversion attitude may be looked at as having too many cheese slices, or cheese slices with no holes, that can act as barriers to progress rather than just to accidents. Conversely Kahneman's hindsight rewards for irresponsible risk-taking can be seen as lack of cheese slices, or slices with too many holes in them. It was put nicely by Reason that *"defenses-in-depth are a mixed blessing. One of their unfortunate consequences is that they make systems more opaque to the people who manage and operate them. They can conceal both the occurrence of their errors and their longer term consequences."*

1.6.3 **Specific Versus Generic Lessons**

It is suggested by Flouron (2011) that there are two categories of lessons that can be learned with respect, for example, to the BP oil blowout in the Deepwater Horizon rig. The first relates to narrow or specific lessons, while the second relates to broader ones. The lens of the former category focuses on such an event as a "failure due to blowout or oil spill"; a type of technical failure, whereas the latter focuses on it as a "disaster"; defined as an occurrence inflicting widespread destruction and distress. Through this latter lens the event is considered broadly in a social science, political, systems theory, and management context, as suggested by the seminal work of Turner (1978) and then followed up by Toft and Rynolds (1997). The two categories of lessons can be visualized as concentric circles, ripples in a pond that move outward from the narrow and specific technical, and regulatory failure issues to the broader economic, social, political, and cultural issues. This is in line with the model of describing what learning from failures is, as proposed by Labib and Read (2013), which comprises feedback to design, using advanced techniques for analysis, and extracting interdisciplinary generic lessons—spanning from the specific to the generic.

1.6.4 **High Reliability Theory Versus Normal Accidents Theory**

There are two schools of thought regarding the value of organizational learning. One school proposes the Normal Accident Theory (NAT), a term coined by Perrow (1984), which claims that since complexity reduces resilience, then accidents—which are complex by nature—cannot be predicted or prevented and hence are "normal" and unavoidable. The theory claims that complexity and tight coupling, together with the ineffectiveness of redundancy in preventing accidents, will inevitably lead to susceptibility to disaster. The other school promotes the high reliability organization (HRO) theory proposed by Rochlin et al. (1987), which believes that organizations can contribute significantly to preventing accidents and hence the emphasis here is not on how accidents happen, but what successful organizations do to promote and ensure safety in complex systems. The HRO theorists offer four principles for such organizations, addressing commitment by leaders, redundancy, culture, and learning. Leveson et al. (2009), Saleh et al. (2010), and Rijpma (1997) provide a comprehensive and balanced account about these two schools of thought in terms of their pessimistic (as outlined by the NAT) and optimistic (as outlined by the HRO theory) approaches. Leveson et al. (2009) give a good illustration of the difference between the two theories: *"Does a plane crash mean that NAT is right or does the reduction in plane crashes over time mean that HRO is right?"*

1.6.5 **Reliability Versus Safety**

Safety and reliability are both important goals, but they are not always positively correlated. In fact, they can be conflicting. Complexity means adding more components to the system and hence reduces probability of success.

Here are some examples to illustrate that reliability and safety are not always positively correlated. The design of the Challenger space shuttle's Solid Rocket Boosters (SRB) sealing, for instance, was redundant. Two O-ring seals were designed-in, so that if one would be eroded and therefore not seal properly, it would be backed up by the other. However, what we now know is that both rings were dependent on weather conditions. If these were such that one would fail then it was probable that the other ring would fail as well. Both O-rings failed during the 1986 launch of the Challenger, causing the disaster. Vaughan (1996) and Rijpma (1997) put it differently; a "common-mode failure" (one caused by the unsuitable weather) overrode the redundancy of the two O-rings. Other sources of complexity may have the same effect. If two redundant parts are very close to each other, for instance, both may be damaged by one blow.

The same concept applies to the nuclear accident at Fukushima, where the cooling pumps were designed as redundant in the sense that they had different power supplies and were very reliable that they could withstand very strong earthquakes, which they eventually did. However, it was not expected (in the design specifications) that the earthquake generated tsunami would be at a height such that all the pumps became submerged. Again, one failure mode affected all the otherwise redundant systems.

Therefore, redundancy, which involves adding more components to the system, can act as a buffer which can be an insulation from failures but also an isolation from the outside world. So failures may get to be unnoticed or hidden in the system. An analogy here is having extra stock as a safety buffer between machines in a manufacturing line. In the Japanese concept of Just in Time (JIT) manufacturing, the objective is to minimize work in stock so that failures in a certain manufacturing cell become quickly and easily identifiable.

1.6.6 Concluding Remarks on Theories Related to Learning from Failures

When examining these dualities of different theories, the first impression one gets is that each pair has conflicting views and there is a need to resolve apparent contradictions. However, it then becomes clearer that these theories are often complementary rather than competitive and that understanding both theories enhances learning from failures. It is even claimed that when applied to some cases, they can reach similar conclusions, as proposed by Rijpma (1997) with respect to the apparent conflict between NAT and HRO theories. Hence when learning to learn from failures, we need to learn about the principles underpinning these theories so that a "balanced" approach is adopted.

There are still, however, some gaps in our knowledge in this field and a need for theories to complete the pieces in the jigsaw puzzle. It was outlined by Sorensen (2002) and Pidgeon (1998) that there are four generic key theoretical issues that still need to be addressed in theories related to safety culture: (i) the paradox that culture can act simultaneously as a precondition for safety and an incubator for hazards; (ii) risk versus incompleteness of knowledge; (iii) organizational construction of acceptable risk; and (iv) critical factors for organizational learning from failures. It is believed by the author that these theoretical concerns still have weight despite the existence of some good contributions in the field.

There are five primary types of accident analysis, as defined by Stellman (1998), namely:

1. Analysis and identification of types of accidents occur and where.
2. Analysis with respect to monitoring developments in the incidence of accidents. This type of analysis looks at factors that affect process operations and could lead to accidents and the development of measures for monitoring the effectiveness of preventive activities.
3. Analysis to prioritize initiatives that call for high degrees of risk measurement, which in turn involve calculating the frequency and seriousness of accidents.
4. Analysis to determine how accidents occurred, and especially to establish both the direct and underlying causes.
5. Analysis for elucidation of special areas which have otherwise attracted attention (a sort of rediscovery or control analysis).

1.7 CASE STUDIES

1.7.1 Choice of Case Studies and Research Methodology

The case studies were chosen to demonstrate learning via the proposed framework of feedback to design, using advanced techniques, and deriving interdisciplinary generic lessons. Case studies were also chosen from different industries and, which to some extent, are believed to have had a major effect on the whole industry where the case study had occurred. As mentioned before, the reason for choosing reported case studies of major disasters is based on the arguments that catastrophic failures are well observed and this visibility tends to encourage organizations as well as policy makers, to learn from those incidents.

Accordingly, all the case studies were constructed to follow a certain format. Firstly, there is a description of what happened, presenting the sequence of events that led to the disaster. This is followed by an investigation of the technical and the logical causes of the failure. Then the consequence of the failure is presented. The tools used for the analysis are then presented. Finally, a set of generic lessons, and recommendations for preventing future system failure, is provided.

I was not sure, however, how to arrange the case studies. Firstly, I asked myself whether the chapters should be arranged by sector, by time of occurrence, or alphabetically? Then I thought that since the objective of what we mean by learning is to study generic lessons spanning across different sectors, and since those generic lessons, by definition, should stand the test of time and not be constrained by the timing of the event, I decided to organize the case studies alphabetically, and hence Part Two is titled "A–Z of Disastrous Case Studies." Unfortunately, or in this particular case, we had better say fortunately, the disasters compiled so far do not yet cover the whole span of the alphabet.

The two techniques that have been used in the majority of the case studies are FTA and RBD. They complement each other, because the outcome of the FTA model is used as an input to produce the RBD model. Both techniques are described in Chapter 2. Information provided in each case study has been analyzed and then developed into an FTA and an RBD model, showing how the models were constructed using the information contained in the narrative of the case study. Other case studies have used a prioritization technique called the AHP, and this has been introduced in Chapter 3.

The fault trees used here are more a general logic tree than a strict classical fault tree as traditionally used to analyze failures at equipment level. The reason for this is that a disaster is depicted as a top event. This is normally not strictly the case in the usual fault tree, i.e., a well-defined subset of a sample space, to which a well-defined binary indicator variable can be attached, i.e., 1 for occurrence, 0 for nonoccurrence. The advantage of using the proposed approach is that it offers richness to the model, so that both subjective judgment and objective evaluation measures are taken into account and hence can act as a good mental model.

It is acknowledged that there could be other techniques that could be used for learning from failures. For example, the AHP as demonstrated in analyzing the Concorde aircraft accident (Davidson and Labib, 2003) and the Bhopal disaster (Labib and Champaneri, 2012; Ishizaka and Labib, 2013), so a chapter has been devoted to describing the AHP method. As far as possible, the same tools have been applied to every case in order to demonstrate the usage of such tools across a variety of technologies and hence show their generic applicability. Specifically, through the case studies we will demonstrate the use of simple analytical tools which managers could find useful for supporting the decision-making process and the design of measures that can lead to an improvement in overall safety performance.

1.7.2 **Types of Recommendations**

A classification of different types of recommendation has been outlined by Toft and Rynolds (1997). Broadly, the classification is into (i) technical and (ii) social. The social category has then been further divided into the subcategories: (i) personnel (e.g., training of staff, monitoring inspection, and establishing monitoring committees), (ii) authority (e.g., failure not allowed by decree—examples are a statement such as "The possibility of an explosion must be recognized when designing" (Toft and Rynolds, 1997), and the imposition of legislation to enforce rules, regulations, and procedures), (iii)information (e.g., communication improvement, routines review, and compiling records of events), and (iv) attempted foresight (e.g., recommendations to organizations other than those investigated and new codes of practice).

Throughout generic lessons derived from each case study will be examined in greater depth. Chapter 13 is dedicated to generic lessons which could be enablers for achieving a learning organization status.

1.8 **CRITICAL COMMENTARY SECTION**

It is noted that not everybody agrees with the findings and conclusions of each chapter. So at the end of some of the chapters there is an alternative view, which is not intended to undermine the issues discussed but rather present other perspectives. The ideas included do not necessarily represent the author's own view, but they are worth consideration and worth initiating some reflection and debate.

Chapter 2

Introduction to Failure Analysis Techniques in Reliability Modeling

2.1 INTRODUCTION

Disasters destroy not only lives but also reputations, resources, legitimacy, and trust. By analyzing disasters, we can learn how to reduce the costs of failures, how to prevent repetitions of failures, and how to make failures rarer. Here, we shall investigate reliability engineering techniques which can be employed in the analysis of disasters and the assessment of safety competence, and which can be part of the intellectual toolkit of all reliability, engineering, safety, and senior management personnel.

These topics will be presented in three phases, as indicated in Figure 2.1 pyramid, which is used for three reasons. Firstly, in many reliability books, the pyramids of Giza are quoted as an example of a reliable system, since reliability is defined as the *"ability of an item to perform its required function over a stated period of time."* Secondly, the program is designed to have a broad introductory base covering the relevant theories, followed by the description of more specific tools that can aid the decision analysis process, and finally focusing on a specific case study. At every step, the level of learning is enhanced and more focused. Thirdly, the sides of the pyramid serve as a space to provide examples.

The base of the pyramid in Figure 2.1 is concerned with the taxonomy of the related theories already discussed in Chapter 1. Here, techniques derived from reliability engineering will be presented and discussed.

The two main techniques that have been used in the case studies are fault tree analysis (FTA) and reliability block diagrams (RBD). They are complementary since the outcome of the FTA is used as an input in the production of the RBD. In a classroom or workshop context, students and practitioners of these techniques are then requested to analyze the information provided for each case study and then develop an FTA and an

Learning from Failures. DOI: http://dx.doi.org/10.1016/B978-0-12-416727-8.00002-3

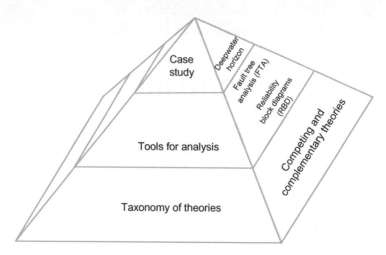

■ **Figure 2.1** Learning from failure chapter outline.

RBD model showing how the models were constructed using the information contained in the narrative of the case study. Students and practitioners are also encouraged to include in their analysis additional techniques, such as failure mode and effects analysis (FMEA), risk priority number (RPN), or iso-critical curves.

2.2 **FTA AND RBD**

It is important here to outline what is meant by FTA and RBD in terms of how they complement each other and how they can be applied to the analysis of disasters (which is slightly different from the way they are normally applied in the analysis of equipment failures).

A fault tree is a logic diagram which shows the relationship between a system failure (i.e., a specific undesirable event in the system) and failures of the components of that system. The undesirable event constitutes the "top event" of the tree and the different component failures constitute the basic events of the tree. For example, for a production process, the top event might be that the process stops, and one of the basic events might be the failure of a particular motor.

There are two important types of logic gate in an FTA, namely, the AND-gate (symbolized by an inverted arc with a horizontal line at the bottom) and the OR-gate (symbolized by an inverted arc with a curve at the bottom), as shown in Figure 2.2.

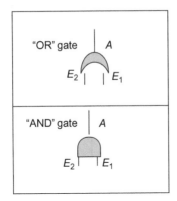

■ **Figure 2.2** The main logical symbols of an FTA: AND- and OR-gates.

An OR-gate indicates that the output event occurs only if one or more of the input events occur. There may be any number of input events to an OR-gate. The implication is that the system involved (plant, process, procedures, and organization) is vulnerable to any one of the inputs, i.e., each one is sufficient to cause the disaster, a situation similar to that proposed by Perrow (1994) in his "normal accident" theory (see, e.g., the Bhopal disaster, which has been summarized as one which was waiting to happen) (Chouhan, 2005; Labib and Read, 2013).

An AND-gate, on the other hand, indicates that the output event occurs only if all the input events occur at the same time. There may be any number of input events to an AND-gate. When modeling a disaster, if the fault tree involved has one gate at the top, i.e., an AND-gate, the implication is that there exists a set of barriers that were all insufficient to prevent the disaster. This is similar to the Swiss cheese model of a disaster proposed by Reason (1997). However, if we are analyzing a new design and there exists an AND-gate at the top, then it is a relatively good, safe, design since many things need to go wrong for the system to fail.

The steps in constructing an FTA are shown in Figure 2.3.

The basic events (see Figure 2.3) are considered to be the "leaves" of the tree, the "initiators," or "root causes." Here it is important to differentiate carefully between the concept of root cause for machines or equipment and that for a disaster or accident. In an accident investigation, if a root cause is perceived as, for example, "someone's behavior"

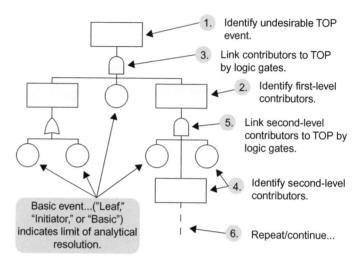

■ **Figure 2.3** Steps for the construction of an FTA.

then it may be likely, as argued by Rasmuseen (1997), that the accident could occur by another cause at another time. Such a root cause is superficial and should be regarded as a symptom rather than the real root cause which needs to be plan- and policy-related with respect to the current status quo ideally should lead to initiation or modification of standard operating procedures (SOPs). Also, a root cause needs to contribute to the three features of how learning from failures is defined, as outlined by Labib and Read (2013), who argue that learning from failures means feedback to design of existing procedures, use of advanced techniques to analyze failures, and generation of interdisciplinary generic lessons.

An RBD is a complementary model of the FTA and is usually drawn after an FTA is constructed. The basic idea is to transform every OR-gate into a series structure and every AND-gate into a parallel one, as shown in Figure 2.4.

For example, in an OR-gate logic, one would imagine items 1, 2, and 3 in Figure 2.4 as valves in a flow line. The flow will cease (in this case the undesirable top event in the FTA) if either valve 1, or 2, or 3 malfunctions and hence an OR-gate can be mapped as a series structure (as shown in the top half of the figure); whereas in the case of an AND-gate if any two valves were malfunctioning we can still bypass the fluid through the third one which is still working, so to stop the flow, all three

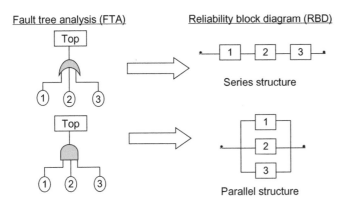

Figure 2.4 The mapping of an FTA to an RBD.

valves need to be malfunctioning and hence an AND-gate in FTA is mapped as a parallel structure in the RBD.

This basic logic may be confusing to those who have an engineering background, especially with respect to PLCs (programmable logic controllers), because for PLCs every series structure in the rungs of the ladder diagram is mapped as AND logic and every parallel structure is mapped as OR. Why is this the case? Because in PLCs we are asking the question "how can this work?," whereas in an FTA we ask "how can this fail to work?" and hence the logic is different.

Therefore, there are several "*golden rules*" that can help in constructing both an FTA and an RBD.

> *Golden Rule No 1: Every OR-gate in an FTA is equivalent to a series structure in an RBD.*
> *Golden Rule No 2: Every AND-gate in an FTA is equivalent to a parallel structure in an RBD.*

When trying to calculate the reliability R_s of the whole system, given that we know the reliabilities R_i of each component, then the following equations apply, as shown in Figure 2.5.

The basic idea is that: $R + F = 1$, as a system has a probability of being either in a reliable (success) or in a fault state. So, in the equation of a series structure (the outcome of an OR-gate), we have

$$R_{sys} = R_1 \times R_2 \times \cdots \times R_n$$

On the other hand, redundancy means adding more components which reduces the probability F of failure and hence enhances the reliability R,

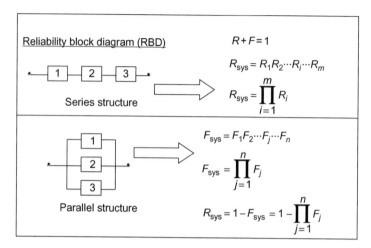

Reliability block diagram (RBD)

$R + F = 1$

$R_{sys} = R_1 R_2 \cdots R_i \cdots R_m$

$$R_{sys} = \prod_{i=1}^{m} R_i$$

Series structure

$F_{sys} = F_1 F_2 \cdots F_j \cdots F_n$

$$F_{sys} = \prod_{j=1}^{n} F_j$$

Parallel structure

$$R_{sys} = 1 - F_{sys} = 1 - \prod_{j=1}^{n} F_j$$

■ **Figure 2.5** The calculations of parallel and series structures in an RBD.

so for a parallel structure all redundant components need to fail for the system to fail and hence

$$F_{sys} = F_1 \times F_2 \times \cdots \times F_n$$

Therefore, reliability increases geometrically with the number of redundant (parallel) parts while cost increases linearly. So when reliability is important, redundant designs become cost effective.

> *Golden Rule No 3: The number of basic events (circles) in the FTA is equal to the number of boxes in the RBD.*

For example, in Figure 2.6, we have three circles and hence the number of "boxes" in the equivalent RBD is also three. Note here that since there is no flow of time from one circle to another (in the FTA), then the sequence of boxes in the RBD does not matter. So boxes 2 and 3 can be swapped, and also in the RBD box 1 can feed to boxes 2 and 3, or we can start from having parallel boxes of 2 and 3 that are on the left and both feed box 1 on the right.

> *Golden Rule No 4: Order (sequence) does not matter.*

The example below shows how an engine can fail to start. Note here that in order to draw the equivalent RBD, it is much easier to start from the top of the tree. This is especially useful when dealing with a complicated hierarchical structure.

> *Golden Rule No 5: Start from the top of the FTA when attempting to model the equivalent RBD (Figures 2.7 and 2.8).*

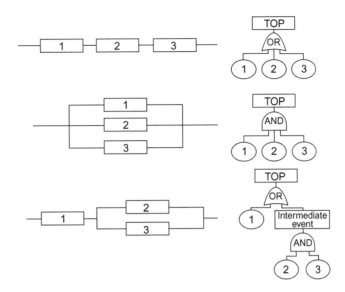

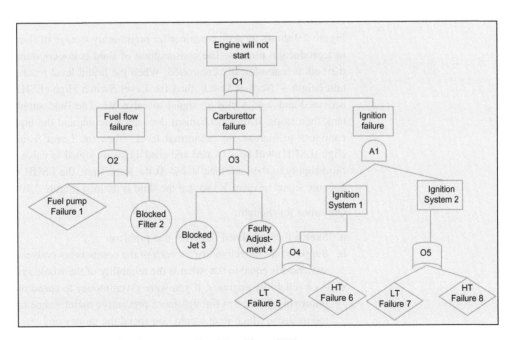

■ **Figure 2.6** FTA and RBD showing the number of circles in an FTA is equal to the number of boxes in an RBD.

■ **Figure 2.7** FTA of an aircraft internal combustion engine. *Adapted from O'Connor (1990).*

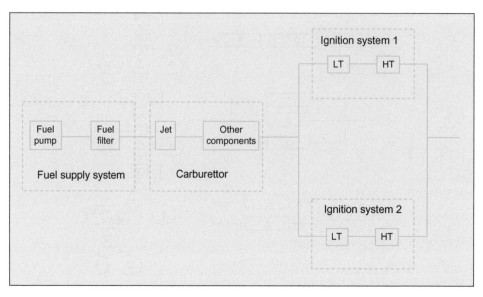

■ **Figure 2.8** Equivalent RBD of an aircraft internal combustion engine. *Adapted from O'Connor (1990).*

2.3 **EXAMPLE: A STORAGE TANK**

Figure 2.9 shows an open container for preliminary storage of fluid for use in a production process. The consumption of fluid is not constant. Filling the tank is automatically controlled. When the liquid level reaches a certain height—"Normal Level" then the Level Switch High (LSH) will be activated and send a closure signal to valve V1. The fluid supply to the tank then stops. If this mechanism does not function and the liquid level continues to increase to "abnormal level," then the Level Switch High High (LSHH) will be activated and send a closure signal to valve V2. The fluid supply to the tank then stops. At the same time, the LSHH sends an opening signal to valve V3 so that the fluid is drained (Figure 2.10).

Questions for thought:

a. Sketch the associated RBD for this problem.
b. Suppose that the reliability for each of the components (valves or limit switches) is equal to 0.8, what is the reliability of the whole system?
c. As a reliability engineer, if you were given money to spend on improving the design through more preventive maintenance or more redundancy, which item would you spend the money on?

The answer to Question (a) is shown in Figure 2.11.

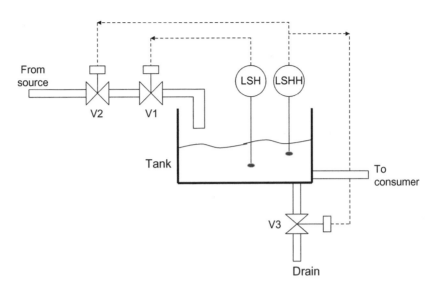

■ **Figure 2.9** Example: Storage tank.

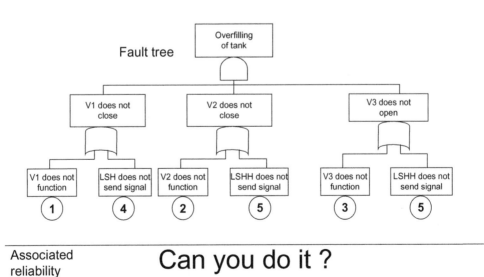

■ **Figure 2.10** The FTA of the storage tank problem.

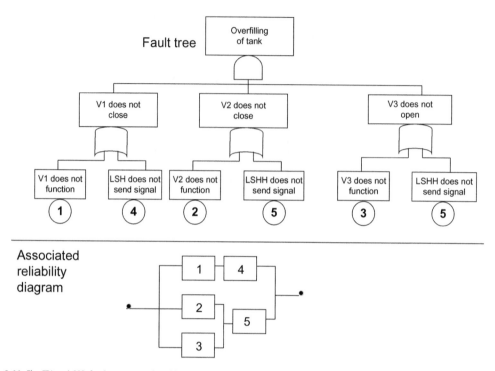

■ **Figure 2.11** The FTA and RBD for the storage tank problem.

As for Question (b), the reliability of the whole system, for series and parallel systems (R_s and R_p):

$$R_s = \prod_{i=1}^{n} R_i$$
$$R_p = 1 - \prod_{i=1}^{m} R_i$$

For each of the three branches:

$$R_I = (0.8)^2 = 0.64$$
$$R_{II} = (0.8)^2 = 0.64$$
$$R_{III} = (0.8)^2 = 0.64$$
$$R_T = [1 - (1 - R_I)(1 - R_{II})(1 - R_{III})]$$
$$R_{II} = [1 - 0.36^3] = 0.95$$

Regarding Question (c), as a reliability engineer, if you are given money to spend on improving the design through more preventive maintenance or more redundancy, you would spend the money on improving box 5 because it controls two boxes (2 and 4). This box is related to the signal from LSHH. Hence more preventive maintenance would be recommended for this signal, or even better, a redesign would be targeted toward separating the signal into two different signals rather than relying on just one signal to control two valves.

So, here is a summary of the Golden Rules when constructing FTAs and RBDs:

1. Every **OR** in an FTA is a **Series** configuration in the equivalent RBD.
2. Every **AND** in an FTA is a **Parallel** configuration in the equivalent RBD.
3. Start from the **Top** of the Tree.
4. Only model **Basic events**.
5. The **order** in an RBD does NOT matter.
6. Look for a **real** root cause.
7. Both FTA and RBD are mental models for **risk analysis** rather than risk assessment.

Although it can be said that FTAs and RBDs are complementary, since the outcome of an FTA is used as an input to the production of an RBD, it is acknowledged that this may be questionable in some cases. Usually, FTA and RBD are alternative techniques, in that FTAs compute failure probabilities while RBDs compute reliability, which is the probability of not having failures.

Moreover, considering that an AND-gate in an FTA is equivalent to a parallel arrangement of blocks in an RBD, and an OR-gate is equivalent to a series of blocks, explains why RBD and FTA are more equivalent to alternative tools than complementary ones. Hence there is a view that they can be used to compute and express similar things, and the proof is that the RBD can be drawn by starting from the FTA and mainly describes the same thing even if from a different viewpoint. This may help to visualize things differently but should lead not to more confusion but rather to a better comprehension of the problem. Our response to this argument, and our choosing to use both techniques, is based on the fact that an RBD clearly shows areas of vulnerability when nodes are arranged in a series structure, whereas relatively safe areas are highlighted when nodes are in parallel. This is not easily seen from the FTA, where it is difficult to arrive at those conclusions from a hierarchical structure of OR- and AND-gates.

It can also be argued that the two techniques are synergistic rather than redundant, as one cannot draw an RBD unless an FTA is constructed first. Here we emphasize that an RBD is different from a process layout, which can be drawn directly; it is more about the root causes that led to the failure mode at the apex of the fault tree, and hence the RBD is an outcome of the FTA. Moreover, both techniques produce different outputs. For example, if data is available about each failure event, one can then derive the total reliability of the system from the RBD, but this derivation of total system reliability cannot be arrived at from the FTA. So, to summarize, an FTA provides the know-how about the logical cause of the failure (hence the use of logic gates), whereas an RBD identifies areas of vulnerability and is capable of calculating total reliability given data about reliabilities of individual events.

2.4 A SIMPLE ILLUSTRATIVE APPLICATION CASE (A CAR ACCIDENT)

Figure 2.12 shows an unfortunate accident in which a car crashes through the wall of a pharmacy and where the driver has literally followed the sign above it!

The first thing to do is to construct all the categories of factors that can cause a car accident. In this case, the factors chosen were related to the conditions of the driver, road, weather, and vehicle. Each was then subdivided into contributing events. For example, with respect to the weather, both rain and poor visibility may together contribute to the accident and the AND-gate indicates that just one of them, on its own, may not be sufficient to cause the accident, whereas with respect to vehicle conditions, it would be argued that if the tire is punctured or there is a mechanical failure then the accident may occur, and hence the use, at that point, of

■ **Figure 2.12** Car accident.

an OR-gate. Obviously, the basic events could be expanded into further ones, but let us assume that we stop here, as shown in Figure 2.13.

In order to construct the equivalent RBD, we apply the golden rules and arrive at Figure 2.14.

Given that the statistical data indicate that accidents tend to be more due to either weather or road conditions rather than to driver or vehicle conditions, one can then assume that $R1$ to $R2 = 0.9$, $R3$ to $R5 = 0.7$, $R6$ to $R7 = 0.7$, and $R8$ to $R10 = 0.9$ (Figure 2.15).

Hence

$$R_{\text{sys}} = [1-(1-R1)(1-R2)][1-(1-R3)(1-R4)(1-R5)][1-(1-R6)(1-R7)]$$
$$\times [1-(1-R8)(1-R9)]R10$$

$$R_{\text{sys}} = [1-(0.1\times0.1)][1-(0.3\times0.3\times0.3)][1-(0.3\times0.3)][1-(0.1\times0.1)]$$
$$\times 0.9 = 0.781 = 78.1\%$$

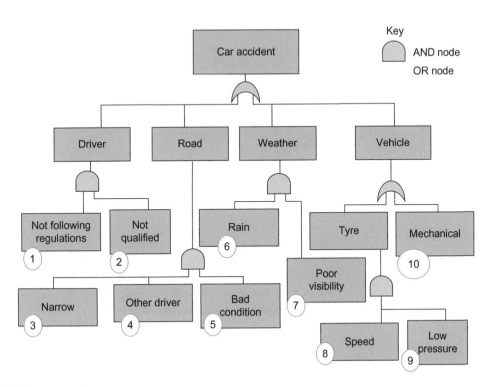

■ **Figure 2.13** FTA of a car accident.

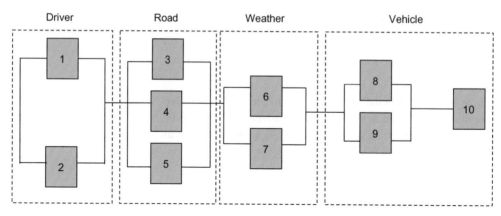

■ **Figure 2.14** RBD of the car accident.

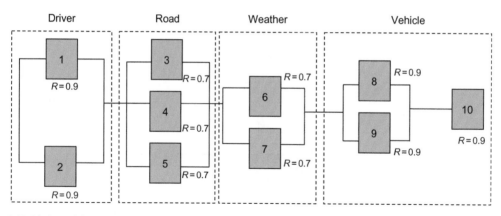

■ **Figure 2.15** Calculation of the system's reliability of the car accident.

Thus, for this case, we should conclude that the likelihood of the car crashing through the window is of the order of $(100 - 78.1)\% = 21.9\%$. Or if the car found itself in these circumstances, there would be about one in five chance of the crash occurring. (Note: the data assumed are purely hypothesized, for illustrative purpose only.)

Chapter 3

Introduction to the Analytic Hierarchy Process

3.1 INTRODUCTION

The analytic hierarchy process (AHP) is a multicriteria decision-making (MCDM) method that helps the decision maker facing complex problems with multiple conflicting and subjective criteria (e.g., problems such as location or investment selection, project ranking, and so forth). Several papers have compiled AHP success stories in very different fields (Zahedi, 1986; Golden et al., 1989; Shim, 1989; Vargas, 1990; Saaty and Forman, 1992; Forman and Gass, 2001; Kumar and Vaidya, 2006; Omkarprasad and Sushil, 2006; Ho, 2008; Liberatore and Nydick, 2008). The oldest reference would appear to be to the work of Saaty (1972), who also gave a precise description of the method in the *Journal of Mathematical Psychology* (Saaty, 1977).

I have used AHP in a variety of applications, including learning from failures, as applied to the Concorde accident (Davidson and Labib, 2003); prioritization of factors in a hybrid integrated approach, applied to analysis of the Bhopal disaster (Ishizaka and Labib, 2013); formulation of a university strategy (Labib et al., 2013); supplier selection (Labib, 2011); motorway planning (Vassou et al., 2006); maintenance policy selection (Labib, 2004; Zainudeen and Labib, 2011); selection of manufacturing systems (Abdi and Labib, 2003); selection of manufacturing paradigms (Alvi and Labib, 2001); prioritization of improvement decisions in manufacturing (Labib and Shah, 2001); developing an intelligent maintenance system (Labib et al., 1998a,b); criticality of machines and their failures (Labib, 1998); formulation of maintenance strategy (Labib et al., 1996, 1997).

Learning from Failures. DOI: http://dx.doi.org/10.1016/B978-0-12-416727-8.00003-5

3.2 AN OVERVIEW OF THE ANALYTICAL HIERARCHY PROCESS

The AHP is an MCDM method that helps a decision-making organization when it is confronting a complex problem which has multiple conflicting and subjective criteria, such as a location or investment selection or a project ranking (see Kumar and Vaidya (2006) and Omkarprasad and Sushil (2006) for reviews of applications of AHP and Ishizaka and Labib (2009) for a review of different methods in assessing priorities). AHP can accommodate the views of a number of decision makers (actors) and the trade-off of their objectives. First introduced by Saaty (1972, 1977, 1980), it requires them to provide judgments about the relative importance of each criterion and then specify a preference on each criterion for each decision alternative.

The first step in AHP is to define the problem to be solved (the goal) and the decomposition of the problem into a decision hierarchy (Vassoulla et al., 2006), as illustrated in Figure 3.1.

The next step is to employ a pair-wise comparison of the criteria among themselves with respect to the goal, as well as between the alternatives with respect to each criterion, in order to establish priorities among the elements in the hierarchy. These comparisons are carried out using Saaty's (1980) predefined one-to-nine ratio scale.

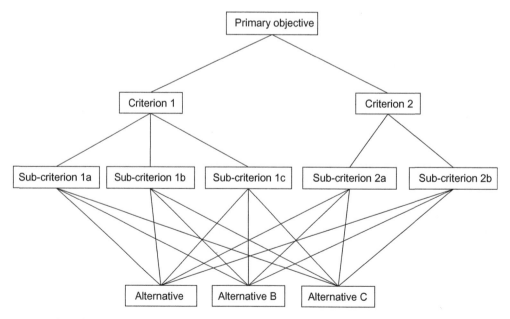

■ **Figure 3.1** A typical AHP decision hierarchy.

The above is then followed by estimation of the relative weights of the elements in each level of the hierarchical model, and computation of the value of the global priorities of alternatives (optional choices) and calculation of consistency (or inconsistency). Finally, sensitivity ("what-if") analysis can be carried out, studying the effect of changing the weights of criteria on the final choice.

In detail, the steps in performing AHP as a method for MCDM are as follows (Labib et al., 2013):

1. Establish the decision context.
   i. Establish the aims of the MCDM.
   ii. Identify the key stakeholders (decision makers and other key players).
2. Identify objectives and criteria.
   i. Identify criteria for assessing the consequences of each option.
   ii. Organize the criteria by clustering them, under high-level and lower-level objectives, in a hierarchy.
3. Identify the options to be appraised (the alternatives).
4. "Scoring." Assess the expected performance of each option against the criteria.
   i. Score the options against the criteria.
   ii. Check the consistency of the scores on each criterion.
5. "Weighting." Assign weights for each of the criteria to reflect their relative importance to the decision.
6. Combine the weights and scores for each option to derive an overall value.
   i. Calculate overall weighted scores at each level in the hierarchy.
   ii. Calculate overall weighted scores.
7. Examine the results.
8. Carry out a sensitivity analysis ("what-if" analysis).
   i. Conduct a sensitivity analysis. Do other preferences or weights affect the overall ordering of the options?
   ii. Look at the advantage and disadvantages of selected options and compare pairs of options.
   iii. Repeat the above steps until a "requisite" model is obtained.

One of the most practical issues in AHP is that it allows for nonconsistent pair-wise comparisons. In practice, particularly with multiple decision makers, perfect consistency is unusual. The pair-wise comparisons in a judgment matrix are considered to be adequate if the corresponding consistency ratio (CR) is less than 10% (Saaty, 1980). The consistency measure is a feedback to the decision maker that helps him/her to capture logical and reasonable preferences when making judgments. It is also a validation because it supports empirical research, conducted by either practitioners or academic researchers, to ensure that questionnaires are not poorly answered.

After the alternatives have been compared with each other, in terms of each one of the decision criteria, and the individual priority vectors have been derived, the priority vectors become the columns of the decision matrix. The weights of importance of the criteria are also determined using pair-wise comparisons. Therefore, given there is one goal, m criteria and n alternatives, the decision maker will create one ($m \times m$) matrix for the criteria and a number of m matrices each of the dimension of ($n \times n$) for the alternatives. The ($n \times n$) matrices will contain the results of $n(n-1)/2$ pair-wise comparisons between the alternatives. Finally, given a decision matrix, the final priorities, denoted by A^i_{AHP}, of the alternatives, in terms of all the criteria combined are determined according to the formula:

$$A^i_{\text{AHP}} = \sum_{j=1}^{n} a_{ij} w_j, \quad \text{for } i = 1, 2, 3, \ldots, m \tag{3.1}$$

There are three outputs that can be produced from the AHP process, namely:

1. An overall ranking, which helps in understanding how each customer is compared to the others.
2. A measure of the overall consistency of the decision maker's preferences, which is a useful feedback for validation of consistency, as explained. Overall inconsistency of less than 10% is normally acceptable as a measure of consistent preferences;
3. A facility to perform sensitivity analysis, which provides information about the causal relationships between the different factors. This can help us to explain and predict the different relationships between criteria and alternatives, and is particularly valuable in creating scenarios for movement in relationships (positive or negative).

In short, AHP is based on four steps: problem modeling, weights valuation, weights aggregation, and sensitivity analysis—a process that will be illustrated, and reviewed, via the solution of a simple problem: the selection of a car to buy.

3.2.1 **Problem Modeling**

As with all decision-making processes, the facilitator will sit a long time with the decision maker(s) to structure the problem, which can be divided

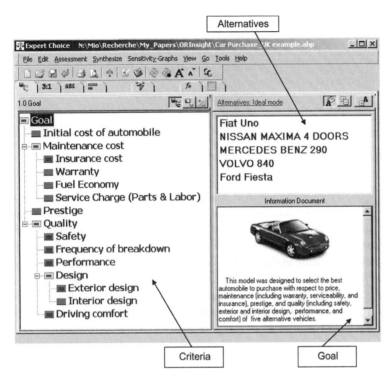

■ **Figure 3.2** Example of hierarchy.

into three parts: the goal (to buy a car), criteria (initial cost, maintenance cost, prestige, quality, and its subcriteria), and alternatives (Fiat Uno, Nissan Maxima 4 Doors, Mercedes Benz 290, Volvo 840, Ford Fiesta) (Figure 3.2). AHP has the advantage of permitting a hierarchic structure of the criteria, which provides users with a better focus on specific criteria and subcriteria when allocating the weights.

Note that in Figure 3.2, the hierarchy is oriented sideways, from left to right, where the goal and criteria, and its subcriteria, are located at the left of the screen, whereas the alternatives (options to choose) are located at the top right side of the screen. The bottom right information is just a further explanation of the nature of the decision-making problem.

3.2.2 **Pair-wise Comparisons**

At each node of the hierarchy, a matrix will collect the pair-wise comparisons of the decision maker (Figure 3.3).

Compare the relative importance with respect to: Goal				
	Initial cost	Maintenance	Prestige	Quality
Initial cost of automobile				
Maintenance cost				
Prestige				
Quality				

■ **Figure 3.3** Comparison matrix of the first node.

Psychologists argue that it is easier and more accurate to express one's opinion on only two alternatives than simultaneously on all the alternatives. This also allows consistency, cross-checking between the different pair-wise comparisons (see Section 3.2.5). AHP uses a ratio scale, which requires no units in the comparison. The judgment is a relative value, or a quotient a/b of two quantities a and b having the same units (intensity, meters, utility, etc.). The decision maker does not need to give a numerical judgment; instead a relative verbal appreciation, more familiar in our daily lives, is sufficient.

If the matrix is perfectly consistent, the transitivity rule holds for all comparisons a_{ij}, i.e.,

$$a_{ij} = a_{ik} \cdot a_{kj} \tag{3.2}$$

For example, if Team A beats Team B two–zero and Team B beats Team C three–zero, then it is expected with the transitivity rule that team A beats team C six–zero ($3 \cdot 2 = 6$). However, this is seldom the case because our world is inconsistent by nature. As a minimal consistency is required to derive meaningful priorities, a test must be done (see Section 3.2.5).

3.2.3 **Judgment Scales**

One of AHP's strengths is the possibility of evaluating quantitative as well as qualitative criteria and alternatives on the same preference scale of nine levels, which can be numerical (Figure 3.4), verbal (Figure 3.5), or graphical (Figure 3.6).

3.2.4 **Priorities Derivation**

Once the comparisons matrices are filled, priorities can be calculated. The first step here is to establish priorities among the elements in the

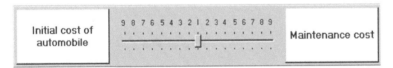

■ **Figure 3.4** Numerical scale.

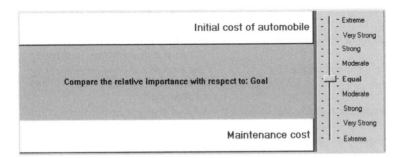

■ **Figure 3.5** Verbal scale.

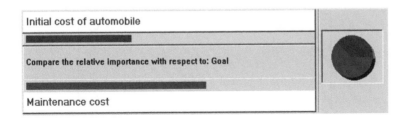

■ **Figure 3.6** Graphical scale.

hierarchy by making pair-wise comparisons of the criteria and alternatives. Given Criterion i and Criterion j, these comparisons are carried out using Saaty's (1980) predefined one-to-nine ratio scale as listed in Table 3.1.

The pair-wise comparisons are carried out for all the factors to be considered and the $n \times n$ positive reciprocal matrix A is generated for $C_1, \ldots, C_n$, $n \geq 2$ criteria, with elements α_{ij}, indicating the value of the criterion i relative to criterion j. Where $\alpha_{ii} = 1$ (i.e., on the diagonal) and if $\alpha_{ij} = k$, then $\alpha_{ji} = 1/k$ (reciprocity property).

Table 3.1 Scale of Relative Importance

Intensity of Importance	Value Description	Explanation
1	*Criterion i* and *Criterion j* are of equal importance	Two activities contribute equally to the objective
3	*Criterion i* is weakly more important than *j*	Experience and judgment slightly favor one activity over another
5	*Criterion i* is strongly more important than *j*	Experience and judgment strongly favor one activity over another
7	*Criterion i* is very strongly more important than *j*	An activity is strongly favored and its dominance demonstrated in practice
9	*Criterion i* is absolutely more important than *j*	The evidence favoring one activity over another is of the highest possible order of affirmation
2, 4, 6, 8	Intermediate values between the two adjacent values	When a compromise in judgment is needed
Reciprocal of above (nonzero)	If *criterion i* has one of the above nonzero numbers assigned to it when compared with *criterion j*, then *j* has the reciprocal value when compared with *i*.	

Source: Saaty (1980).

	C_1	...	C_j	...	C_n
C_1	α_{11}		α_{1j}		α_{1n}
		α_{ii}	α_{ij}	α_{ik}	
C_j	α_{ji}	α_{ji}	α_{jj}	α_{jk}	α_{jn}
C_n	α_{n1}		α_{nj}		α_{nn}

Approximating the weight vector of the above matrix A, with n objectives, takes the form illustrated below. Where $w_i > 0$, $i = 1, \ldots, n$, denotes the weight of objective i. The next step is the calculation of a list of the relative weights of the criteria under consideration. This requires normalization of each column j in A such that $\sum_j a_{ij} = 1$.

	C_1	C_2	...	C_n
C_1	$\frac{w_1}{w_1}$	$\frac{w_1}{w_2}$		$\frac{w_1}{w_n}$
C_2	$\frac{w_2}{w_1}$			
C_n	$\frac{w_n}{w_1}$	$\frac{w_n}{w_2}$		$\frac{w_n}{w_n}$

⇓ normalize columns

	C_1	C_2	...	C_n
C_1	w_1	w_1	...	w_1
C_2	w_2			
C_n	w_n	...		w_n

For each row i in the above resulting matrix, the average value is computed, such that

$$w_i = \frac{1}{n} \sum_j a_{ij}$$

where w_i is the weight of criterion i in the weight vector $w = [w_1, w_2, \ldots, w_n]$ recovered from matrix A, with n criteria, by finding a (nontrivial) solution to a set of n equations with n unknowns. This is achieved by solving the eigenvector problem: $Aw = \lambda_{max}w$. The sum of the weights is taken as equal to 1, resulting in a unique nontrivial solution. Where λ_{max} is the principle eigenvalue for the pair-wise comparison matrix A.

One of the most practical features of the AHP methodology is that it allows for slightly nonconsistent pair-wise comparisons. If all the comparisons are perfectly consistent, then the relation $\alpha_{ij} = \alpha_{ik}\alpha_{kj}$ should always be true for any combination of comparisons taken from the judgment matrix.

In practice, however, perfect consistency sparingly occurs. The pair-wise comparisons in a judgment matrix are considered to be adequate if the corresponding CR is less than 10% (Saaty, 1980). First the consistency index (CI) will be estimated by computing Aw and approximating the maximum eigenvalue, λ_{max}, using the following:

$$\lambda_{max} = \frac{1}{n} \sum_{i=1}^{n} \frac{i\text{th entry in } Aw}{i\text{th entry in } w} \tag{3.3}$$

Then, the CI value is calculated by using the expression

$$CI = \frac{\lambda_{max} - n}{n - 1} \tag{3.4}$$

The synthesis step follows, after the alternatives have been compared with each other, in terms of each one of the decision criteria, and the individual priority vectors have been derived. The priority vectors become the columns of the decision matrix (not to be confused with the judgment matrices of the pair-wise comparisons). The weights of importance of the criteria are also determined using pair-wise comparisons. Therefore, given there is one goal, m criteria and n alternatives, the decision maker will create one $m \times m$ matrix for the criteria and m $n \times n$ matrices for the alternatives. The $n \times n$ matrices will contain the results of $n(n-1)/2$ pair-wise comparisons between the alternatives. Finally, given a decision matrix, the final priorities, denoted by A^i_{AHP}, of the alternatives in terms of all the criteria combined are determined according to the formula:

$$A^i_{AHP} = \sum_{j=1}^{n} a_{ij}w_j, \quad \text{for } i = 1, 2, 3, \ldots, m \tag{3.5}$$

Clearly, AHP is most efficiently applied when the total number of criteria and alternatives is not excessive. Several computer software packages are available to perform the AHP calculations, although the user is always obligated to input the pair-wise comparison scores.

3.2.5 Consistency

As priorities make sense only if derived from consistent or near-consistent matrices, a consistency check must be applied. Saaty (1977) has proposed a CI, which is related to the eigenvalue method, i.e.,

$$CI = \frac{\lambda_{max} - n}{n - 1} \tag{3.6}$$

where

$n =$ dimension of the matrix
$\lambda_{max} =$ maximal eigenvalue

If the CR, the ratio of CI and the random index (RI) (the average CI of 500 randomly filled matrices), is less than 10%, the matrix can be considered as having an acceptable consistency.

Saaty (1977) calculated random indices as given in Table 3.2.

3.2.6 Sensitivity Analysis

The last step of the decision process is the sensitivity analysis, where the input data are slightly modified in order to observe the impact on the results. If the ranking does not change, the results are said to be robust. The sensitivity analysis is best performed with an interactive graphical

Table 3.2 Random Indices								
n	**3**	**4**	**5**	**6**	**7**	**8**	**9**	**10**
RI	0.58	0.9	1.12	1.24	1.32	1.41	1.45	1.49
Source: *From Saaty (1977).*								

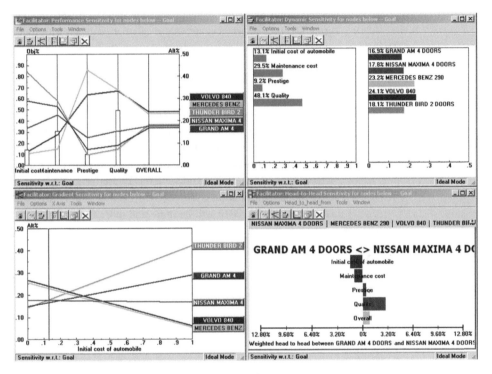

■ **Figure 3.7** An example of four possible graphical sensitivity analyses in Expert Choice.

interface, Expert Choice, which facilitates different sensitivity analyses and various graphical representations (Figure 3.7).

3.3 **CONCLUSION AND FUTURE DEVELOPMENTS**

Decisions that need support methods are difficult by definition and therefore complex to model. A trade-off between the degree of perfection of the modeling and its usability should be achieved. AHP has done this in the past and will continue to be useful for decision problems in the future. In particular, AHP has progressed through the academic community to be widely used by practitioners, which has certainly been due to its ease of applicability and to its structure, which reflects the intuitive way in which managers solve problems. The hierarchical modeling of the problem, the possibility of adopting verbal judgments, and the verification of the consistency are its major assets.

Expert Choice, the user-friendly supporting software, has certainly largely contributed to the success of the method. It incorporates intuitive graphical user interfaces, automatic calculation of priorities and inconsistencies, and several ways to process a sensitivity analysis. Today, several other supporting software packages have been developed: Decision Lens, HIPRE 3 + , RightChoiceDSS, Criterium, EasyMind, WebAHP, etc. not to mention that a template in Excel could also be easily generated. Along with its traditional applications, a new trend, as compiled by the work of Ho (2008), is to use AHP in conjunction with other methods, such as mathematical programming techniques like Linear Programming, Data Envelopment Analysis (DEA), Fuzzy Sets, House of Quality, Genetic Algorithms, Neural Networks, SWOT analysis, and recently with Geographic Information System (GIS) (Temiz and Tecim, 2009). There is little doubt that AHP will be more and more frequently used.

A—Z of Disastrous Case Studies

Chapter 4

Bhopal Disaster—Learning from Failures and Evaluating Risk

4.1 INTRODUCTION

This chapter is based on two analyses of the causes of the Bhopal incident carried out by the author and colleagues. The first study (Labib and Champaneri, 2012) provides an overview of the incident, uses fault tree (FTA), reliability block diagram (RBD), and failure mode and effect analyses, and includes models, developed by students on the author's courses in reliability engineering, which show how different mental maps can be created for the same narrative of the disaster. The second study (Ishizaka and Labib, 2013) uses much of the analysis developed in the first work but also introduces a new concept of integrating and AHP (analytical hierarchy process) (and proposed a new logic gate in the FTA).

4.2 BHOPAL NARRATIVE AND THE INCORPORATION OF FTA AND RBD

This chapter aims to identify the economic, technological, and organizational errors constituting the root causes of the Bhopal disaster of December 2 and 3, 1984. The technical causes of the failure, from a design and operations point perspective, are highlighted. An investigation is then carried out to determine the major consequences of the failure. FTA and RBD analysis are used to model the evolution of the accident and its probability of occurrence. The innovative aspect of this study is that whereas we traditionally use such an analysis at an equipment level here we apply it to analyze a catastrophic event. Recommendations in terms of emergency and contingency planning are then provided. It is concluded that, in multinational companies (MNCs) projects, designs of installations need to be peer reviewed, more stringent environmental, health, and safety considerations adopted, and governments need to be aware of the requirement for segregation of hazardous operations from other facilities and from adjacent domestic populations.

Learning from Failures. DOI: http://dx.doi.org/10.1016/B978-0-12-416727-8.00004-7

4.2.1 **Introduction**

In 1984, Bhopal city, located in the center of India with a population of approximately 1.4 million, became one of the best known places in the world—but for all the wrong reasons. On December 3, 1984, when the town's people slept, the Union Carbide (UC) pesticide plant (4.8 miles away) unleashed "hell on earth." Poisonous gases were released and killed some 3000 people. These gases included a gas used in the First World War that attacks the "wet" parts of our bodies, such as the eyes, mouth, and throat. This then enters the lungs, where it reacts with bodily fluids, filling the lungs, and drowning "from the inside." This was a disaster the town might eventually (over a long period of time and with help) have come to terms with, were it not for the following facts:

- The deaths did not stop at 3000 (there are reputed to have been some 20,000 to date) and still, to this day, approximately 200,000 people are suffering from serious ill health issues as a result.
- "It was an accident waiting to happen." By comparison with similar plants in the United States and India, the Bhopal plant was neglected to say the least (Chouhan, 2005). Cost cutting measures were introduced at the cost of safety. The value of human life in India was not a priority.
- The lies—Management at the plant (none of whom died that night) commented that the gas was similar to tear gas and said the effects would fade in 3 days . . . some 20 years later the effects are still evident. The Union Carbide Corporation and the Indian Government claimed, until 1994, that the gas methyl isocyanate (formally known as MIC, and registered trademark "Sevin") had no long-term effects.
- Since 1976, a considerable history of events of the sort that contributed to the eventual 1984 disaster, had gone unheeded by the Corporation, and to this day they have not claimed full responsibility for any wrong doing; nor does anyone sit in jail for the "murder" of so many.
- Reports issued months prior to the incident by scientists within Union Carbide Corporation warned of the possibility of an accident almost identical to that which happened; reports were ignored and never presented to senior staff.

In this chapter, an attempt has been made to develop an objective fault tree which can reveal what can be learned from this terrible incident. It "was an accident waiting to happen." This chapter intends to show this by:

- Discovering the technical causes of the failure from a design and operations perspective.
- Identifying the major consequences of the failure—then and today.

- Using an FTA and RBD analysis to determine the probability of such an occurrence happening.
- Recreating a "new" FTA by implementing the minimum cut set method.
- Setting recommendations regarding emergency and contingency planning.

4.2.1.1 *Background*

In 1969, UC, a multinational corporation, established a small Indian subsidiary, Union Carbide India Ltd (UCIL) to manufacture pesticides at Bhopal in India. The Indian plant offered the competitive advantages of low labor costs, access to an established and rapidly growing market, and lower operating costs. In addition, UCIL was able to exploit the country's lax environmental and safety regulations as India strived to attract large MNCs for its developing industrialization program. Until 1979, UCIL imported MIC, a key component in the production of pesticides, from its parent company, UC in the United States. The new Bhopal facility was advertised as being designed and built on the basis of 20 years' experience from UC's MIC plant in West Virginia, USA.

4.2.1.2 *Installation*

As early as 1972 a UC internal report had recommended that if additional MIC plants were to be built, they should be constructed of materials as good as those used in the West Virginia plant. It became clearly evident that although UC engineers oversaw the design, build and operation until the end of 1982, along with technical support and safety reviews, the Indian facility underwent cost cutting programs in design and construction which were not mirrored in comparable Western plants:

1. Carbon steel pipe, which is more corrosive, replaced stainless steel piping.
2. The number and quality of safety devices was reduced (a saving of $3–6 million).
3. In Western plants, installed safety devices were automatically controlled with backup devices—at Bhopal they were manually controlled.
4. At similar Western plants, computerized early warning systems sensed leaks, monitored their rates and concentrations, and were linked to a telephone system to automatically dial out alerts. At Bhopal there weren't even any emergency planning measures.

5. At Bhopal a single vent gas scrubber (VGS) was installed. The equivalent plant in the United States had four such scrubbers, i.e., a highly reliable, multiple redundancy arrangement.
6. At Bhopal only one flare tower was installed, i.e., there was no redundancy. The equivalent plant in the United States had two.
7. In Bhopal, there was no unit storage tank between MIC manufacture and the large storage tank, which should be installed to check for purity. This was designed in and installed on the US plant.

Due to design inadequacies, none of the six main safety features of the plant was efficient, but also on the night of the incident none was operational (due to a pressurized maintenance schedule caused by understaffing).

At the local level, no emergency planning was undertaken prior to the commissioning of the plant. In the United States, emergency planning was mandatory and involved all of the emergency services and a public broadcasting system.

Prior to the disaster, operating incidents resulting in plant workers being killed or injured, along with minor releases of toxic gases, had caused UC to send a team of US experts to inspect the Bhopal plant as part of a safety inspection in May 1982. The report, which was passed to UC's management in the United States, indicated that there was "*a serious potential for sizeable releases of toxic materials in the MIC, unit either due to equipment failure, operating problems, or maintenance problems,*" thus requiring various changes to reduce the danger of the plant; there is no evidence that the recommendations were ever implemented (Weir, 1987).

4.2.1.3 *Precursors Leading to the Disaster*

Prior to the disaster, training, manning levels and the educational standards of the employees of the plant workforce were reduced. Between 1980 and 1984, the plant's workforce was reduced by half with no clear investment in technology to warrant this reduction.

The basic operation of the plant was further compromised by management decisions to operate the plant either outside its designed operating parameters or by implementation of revised processes to ensure continued production while essential components of the system had known defects; these defects had the potential to impact on the safety integrity of the plant.

4.2.2 **Direct Causes of the Accident**

The production of a deadly cloud of MIC was produced as a consequence of a cheap engineering solution to a known maintenance problem. A "jumper line" connected a relief valve header to a pressure vent header enabling water from a routine washing operation to pass into MIC Storage Tank 610. The ingress of water to the MIC tank created an uncontrollable runaway exothermic reaction. The reaction products passed through the process vent header to the jumper line, to the relief valve vent header, on to the VGS and finally to atmosphere through the atmospheric vent line. The toxic gases were discharged for 2 h 15 min.

The release of toxic gases was assisted by the following defects and lapses in standard operating procedures which, in many instances, could have easily been averted, namely:

- MIC Storage Tank 610 was filled beyond recommended capacity. Functional contents gauges should have provided warning of this and the process halted until rectified.
- A storage tank which was supposed to be held in reserve for excess MIC already contained MIC (Cassels, 1993). The reserve storage tank should have been empty and there should have been a formal requirement that any production should have been halted until this had been established. However, the "hold point" in the control process prior to production was ignored (Chouhan, 2005).
- The blowdown valve of the MIC Storage Tank 610 was known to be malfunctioning; consequentially it was permanently open. This valve should have been repaired or the tank should have been removed from service until repaired.
- The danger alarm siren used for warning the adjacent residential communities was switched off after 5 min in accordance with revised company safety practices. This clearly highlights why the site required emergency procedures to be in place and continually reviewed.
- The plant superintendent did not notify external agencies of the accident and initially denied the accident had occurred. This was clear negligence on behalf of the management but typified the poor health and safety culture within the plant.
- The civic authorities did not know what actions to take in light of there being no emergency procedures in place and were uninformed of the hazardous materials stored within the plant. The requirements for good communications and established emergency procedures with local agencies and emergency services highlighted these shortfalls.
- Gauges measuring temperature and pressure in the various parts of the facility, including the crucial MIC storage tanks, were so notoriously unreliable that workers ignored early signs (Weir, 1987). The company should have had a robust maintenance regime which

should have prevented this, coupled with a safety culture which should have questioned any unsafe conditions.

- The refrigeration unit for keeping MIC at low temperatures, and therefore making it less likely to overheat and expand should contamination enter the tank, had been shut off for sometime (Weir, 1987). This issue could have only been resolved by the management having a commitment to safety and process guarding as opposed to profit generation.

The failings below are attributable to design reductions and the fact that UCIL was able to dilute its safety protection devices in order to maximize profits, while any local peer reviews of designs by local safety/engineers was nonexistent:

- The gas scrubber, designed to neutralize any escaping MIC, had been shut off for maintenance. Even had it been operative, postdisaster inquiries revealed the maximum pressure it could handle was only one-quarter that which was actually reached in the accident (Weir, 1987).
- The flare tower, designed to burn off MIC escaping from the scrubber, was also turned off, waiting for the replacement of a corroded piece of pipe. The tower, however, was inadequately designed for its task, as it was capable of handling only a quarter of the volume of the gas released (Weir, 1987).
- The water curtain, designed to neutralize any remaining gas, was too short to reach the top of the flare tower where the MIC billowed out (Weir, pp. 41–42).
- There was a lack of effective warning systems; the alarm on the storage tank failed to signal the increase in temperature on the night of the disaster (Cassels, 1993).

4.3 **THEORY/CALCULATION**

In this section, FTA and RBDs are used to map the root causes of the disaster and calculate its overall reliability (as an outcome of probability of occurrence).

The RBD can be derived after constructing an FTA. The parallel and series connections in the RBD, which are derived from the AND and OR gates respectively of the FTA, describe how the system functions (or fails to function), but the RBD does not necessarily indicate any actual physical connection or any sequence of operation. In other words, it does not model the flow of material or sequence of time events, but instead models the logic connecting the root causes that led to the failure mode at the apex of the tree in the FTA.

There are many benefits that can come out from an analysis based on FTA and RBD modeling. First, it helps to highlight vulnerable, or weak, areas in the plant that need attention in the form of adding, e.g., built-in-test systems, redundancy, or more preventive maintenance. Second, it acts as a knowledge base indicating how a system fails and hence can be used for diagnostics or fault finding. Finally, given the value of availability for each box in the RBD model—where people usually mistakenly denote it as "R" for reliability although strictly speaking it should be "A" for availability—one can then calculate the whole system's availability. Hence such an analysis is useful for showing how to improve systems reliability by preventing things from going wrong, as well as for showing how to recover the situation by restoring things when they do go wrong.

Normally, we use FTA and RBD to model a failure mode at an equipment level or for a certain machine seen as a system. Such a failure mode may for example be a "certain motor fails to start." However, in this chapter, we shall apply the same methods of analysis but rather on a larger scale where the failure mode is a disastrous situation such as the Bhopal incident. So here our system failure is a disastrous event rather than the failure of a certain item or equipment. Here, there are two distinct features that should be considered. Firstly, in a disastrous situation, we are dealing with a complex situation where there are usually human, social, and environmental factors, where reliability is difficult to evaluate, in play. Secondly, the whole meaning of availability becomes philosophical, and sometimes even confusing, in the context of modeling a disastrous situation. We pose a fundamental question: would one expect a low or a high figure of total availability from a disaster?

To attempt to answer this seemingly simple question, one needs to go back to the fundamental definitions of these terms. Availability is the measure of reliability and is calculated as a function of both mean time between failures (MTBF) and mean time to repair (MTTR), where both MTBF and MTTR are related to severity and frequency of a failure respectively, and hence MTBF is a functional reliability measure, whereas MTTR is a maintainability measure. Since a disaster is by its very nature a severe and yet a rare event, one would normally expect high figures of availability due to its very low frequency (a one-off event). However, this is not the case when the existing design of the system (in terms of its design integrity and operation) is not fit for purpose and hence it is a disaster waiting to happen, and in this situation one

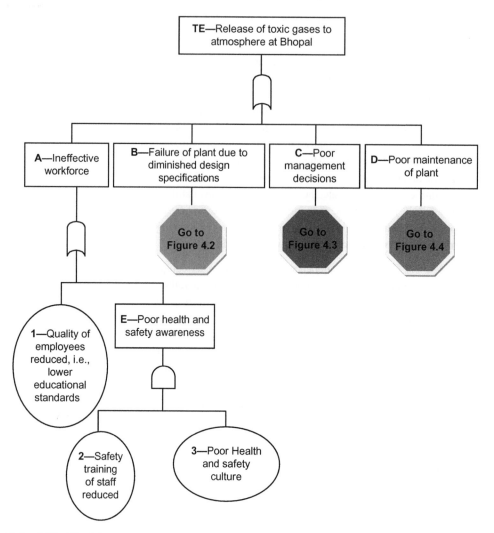

■ **Figure 4.1** Overall FTA of Bhopal disaster.

would expect that total systems availability would be rather low. Figure 4.1 shows the overall FTA of the Bhopal disaster.

Figures 4.2–4.4 show further analysis and extension of Figure 4.1. In Figure 4.2, the fault tree of failure of plant due to diminished design specifications of plant is presented.

Figure 4.3 is a fault tree of management decisions failures.

Figure 4.4 is a fault tree of plant maintenance inadequacies.

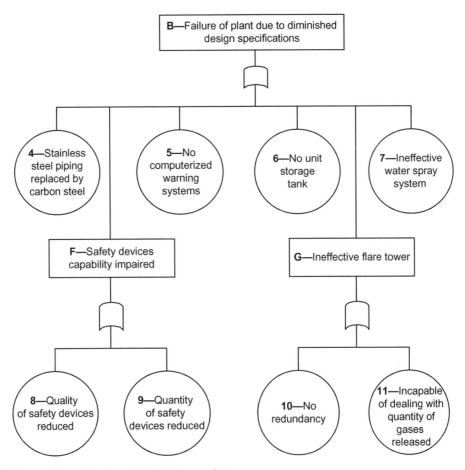

■ **Figure 4.2** Fault tree of failure of plant due to diminished design specifications.

The equivalent RBDs for each of the preceding fault trees are then shown in Figure 4.5.

The combined RBD related to the Bhopal disaster can then be constructed as a combination of all the RBDs presented in Figure 4.5, where they are all linked in a series structure (due to the existence of the OR gate at the top of the tree) which signifies a very vulnerable state "an accident waiting to happen," because in such a series dependency there are many things of which, if any (or all) go wrong could participate into a disaster.

In Table 4.1, estimated probabilities of failures for various events discussed in Section 4.3 are listed. This is a speculative attempt at deriving

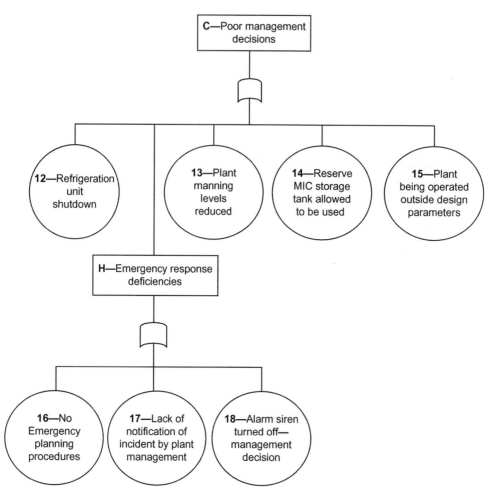

■ **Figure 4.3** Fault tree of management decisions.

the probability of occurrence of the disaster. Using these data and applying straightforward Boolean analysis of the logic of the fault trees can lead to a more accurate estimate, which is beyond the scope of this chapter. Estimates of P_f (probability of system failure) are used as a measure of unreliability where the sum of P_f and P_s (probability of system success) equals one as the system is either in a failed or running states. Again the word "probability" here may have different meanings, as it can mean a measure of confidence or a measure of availability. Anyway, we use it in this context to provide us with an indication of the relative

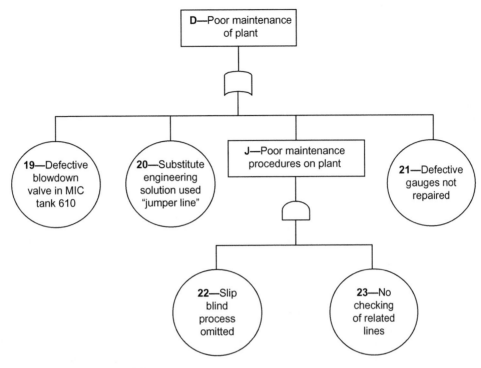

■ **Figure 4.4** Fault tree of poor maintenance of plant.

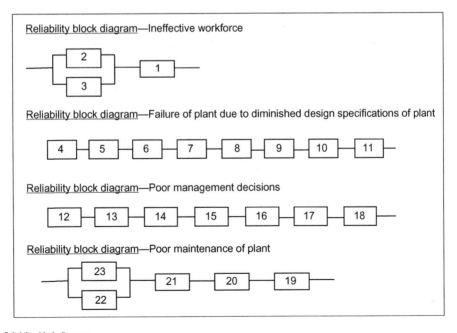

■ **Figure 4.5** Reliability block diagrams.

Table 4.1 Estimated Probabilities of Failures for Various Events

Event No.	Comments	$P_f = F/S + F$	Actual P_f Value
1—Quality of employees reduced	Reduction of operators with high school education over 5 years = 9 No. of operators with high school education at time of disaster = 6 No. of hours for 5-year period = $24 \times 365 \times 5 = 43,800$	$= (9/6 + 15)/43,800$	9.6×10^{-6}
2—Safety training of staff reduced	Original No. of training days = 18 months $\times$ 30 days = 540 Reduction in training days during 5-year period = 523 No. of actual training days = 17 No. of hours for 5-year period = $24 \times 365 \times 5 = 43,800$	$= (523/540 + 17)/43,800$	21.4×10^{-6}
3—Poor health and safety culture	Assumption—health and safety culture dependent upon quality and training of staff P_f value = event 1 + 2	$= 9.6 \times 10^{-6} + 21.4 \times 10^{-6}$	31.1×10^{-6}
4—Stainless steel piping replaced by carbon steel	Guess—No. of new repairs performed = 1000 No. of repairs performed since opening of plant = 25,000	$= 1000/25,000 + 1000$	40×10^{-3}
5—No computerized warning systems—human detection	Guess—No. of failures detected by staff = 20000 No. of failures missed during inspections = 2000	$= 2000/20,000 + 2000$	90.9×10^{-3}
6—No unit storage tank	Guess: No unit storage tank fitted to check purity therefore assuming check performed once a week during 5 years $52 \times 5 = 260$ System would fail once ever week day over 5-year period = $365 \times 5 = 1825$	$= 260/1825$	142.4×10^{-3}
7—Ineffective water spray system	Guess: System would fail to suppress gases due to design error associated with height in MIC area which represented area only 1/5000 of plant System failed in MIC area on day over 5-year period of MIC production = $365 \times 5 = 1825$	$= (1/5000)/1825$	109.6×10^{-9}
8—Quality of safety devices reduced	Assumptions: Reduced quality of safety devices resulted in 50% increase failure rate Guess previous P_f value = 8×10^{-3}		16×10^{-3}
9—Quantity of safety devices reduced	Assumption: Number of devices reduced by 25% therefore P_f value increased by 25% Guess: Previous P_f value = 8×10^{-3}		10×10^{-3}

Item	Description	Calculation	Value
10—No redundancy (flare tower)	Two built in US plant, one installed in India. Item under maintenance for	$(1/2)/43{,}800$	11.4×10^{-6}
11—Incapable of dealing with quantity of gases	Duration of use of single flare tower over 5 years $= 24 \times 365 \times 5 = 43{,}800$ h. System not designed to deal with volume of gases. Therefore, the system failed to handle this volume of gas for 2.5 h (time of disaster). Duration in hours system operating over 5 years $= 24 \times 365 \times 5 = 43{,}800$ h	$2.5/43{,}800$	57.1×10^{-6}
12—Refrigeration unit shutdown	Unit shutdown for past year $= 365$ days. MIC production over last 5 years $= 365 \times 5 = 1825$	$365/1825 + 365$	166.3×10^{-3}
13—Plant manning levels reduced	Overall 20% reduction of staff in 4 years. Duration $= 365 \times 4 = 1460$ days	$(20/100)/1460$	136.9×10^{-6}
14—Reserve MIC storage tank allowed to be used	Assumption: MIC storage tank used for 50% of time. Duration of use $= 5$ years $= 1825$ days	$(50/100)/1825$	273.9×10^{-6}
15—Plant being operated outside design parameters	Assumption: 10% of plant being operated outside design parameters for 5 years	$(10/100)/1825$	54.8×10^{-6}
16—No emergency planning procedures	No emergency planning in place—operation failed on day 5 years of MIC production at plant $= 5 \times 365 = 1825$	$1/1825$	0.55×10^{-3}
17—Lack of notification of incident by plant management	No emergency planning in place—operation failed on day 5 years of MIC production at plant $= 5 \times 365 = 1825$	$1/1825$	0.55×10^{-3}
18—Alarm siren turned off—management decision	No. of hours siren turned off $= 2$. No. of hours available for use $= 24 \times 365 \times 5 = 43{,}800$	$2/43{,}800$	45.7×10^{-6}
19—Defective blowdown valve in MIC tank	Total number of days defective in past 5 years $= 12$. Total number of days in 5 years $= 1825$	$12/1825$	6.6×10^{-3}
20—Substitute engineering solution used "jumper line"	No. of times procedure used $= 150$. No. of flushing operations $= 450$	$150/450 + 150$	250×10^{-3}
21—Defective gauges not repaired	Guess: No. of gauges defective or still in use $= 1320$. Total number of gauges on plant $= 6000$	$1320/1320 + 6000$	180.3×10^{-3}
22—Slip blind process omitted	Guess: No. of procedures requiring slip blinds but not used $= 50$. Total number of procedures requiring slip blinds $= 220$	$50/220 + 50$	185.1×10^{-3}
23—No checking of related lines	Guess: No. of times procedure required during maintenance in 5 years $= 300$. Total number of maintenance procedures in 5 years $= 15{,}000$	$300/15000 + 300$	19.6×10^{-3}

Assumptions—MIC production in Bhopal commenced in 1979, i.e., 5 years before disaster.

importance (priorities) of different factors that could lead to the disaster. Note that the numbers labeling the various FTA events and/or RBD boxes in the figures refer to the numbers used to list the various events/ factors listed in Table 4.1.

4.4 **DISCUSSION**

UCIL had allowed safety standards and maintenance at the plant to deteriorate to dangerous levels even though the potential for such an incident had been highlighted 2 years prior in a UC internal report. Clearly, UCIL had dropped the operating and safety standards of the Bhopal facility well below those maintained in the near identical facility in West Virginia. The fact that UCIL was able to do this was due in part to lack of enforcement by the Indian Government of safety and environmental laws and regulations. Immediately after the disaster in India, UC, while maintaining no knowledge of the cause of the accident in India, shut down its MIC plant in West Virginia to allow $5 million of changes to its safety devices to be accomplished.

4.5 **CONCLUDING REMARKS**

The Indian Government, although keen to attract foreign investment, needed to factor in basic safety requirements for its citizens. During future MNC projects, designs of installations need to be peer reviewed and more stringent environmental, health, and safety measures adopted.

During any future plant builds, standards of materials and equipment used should reflect those used in Western countries. In the light of the Bhopal disaster, MNCs need to be aware that reduction in safety standards as a means of improving profit margins is not an option.

Governments also need to be aware of the requirement for segregation of hazardous operations from other industrial facilities and from adjacent populations. In the case of Bhopal, the local communities and "squatter camps" should have been relocated prior to any company being given permission to start mass production of inherently dangerous material.

A means of guarding operating processes, along with habitual safety checking, needs to be implemented and established as a cornerstone of any safety culture within hazardous plants. Such cultures need to be developed so that questioning attitudes are commended rather than discouraged and safety is the primary driver rather than profit.

MNCs need to reinstigate high levels of safety training to improve employees' awareness of hazards. In addition, the quality and numbers of personnel should not be reduced at the expense of safety in order to bolster company profits. MNCs attracted to third world countries by the prospect of cheap labor and potentially less stringent environmental, health, and safety legislation needs to consider the adverse impact on their business brought about by the focused media coverage resulting from perceived neglect of the health and safety of their workforce, which ultimately impacts on their company reputation.

The main significance of this work is that it demonstrates that learning can be addressed via three perspectives, namely (i) from feedback from the users (maintenance) to design, (ii) by the incorporation of advanced tools in innovative applications, and (iii) by the fostering of interdisciplinary approaches and generic lessons. The basic findings from the present work are therefore related to the feedback process through advice to both future MNC projects regarding designs of installations and recommendations to governments regarding health and safety measures. Here, tools such as FTA, RBDs, and cut set calculations have aided the development of an objective model, which delivers salutary lessons about this terrible incident. The aim has been to develop a generic approach that can be used to learn from any future disasters.

4.6 **CRITICAL COMMENTARY SECTION**

It is noted that not everybody agrees with the findings and conclusions of each chapter. So at the end of some of the chapters there is an alternative view, which is not intended to undermine the issues discussed but rather present other perspectives. The ideas included do not necessarily represent the author's own view, but they are worth consideration and worth initiating some reflection and debate.

CRITICAL COMMENTARY: FAIL SAFE SYSTEMS

People sometimes misconceive what a 'fail-safe' system really is. It is not just another redundant system. Fail safe systems are the outcome of the Japanese approach to continuous improvement (Poke Yoka) to prevent those failures that are due mainly to human mistakes. Chase and Stuart (1994) offer good examples

1. The locks on aircraft or train lavatory doors, which must be turned to switch the light on.

(Continued)

CRITICAL COMMENTARY: FAIL SAFE SYSTEMS (CONTINUED)

2. The height bars on amusement rides to ensure that customers do not exceed height limitations.
3. The beepers on ATM machines to ensure that credit/debit cards are removed
4. The paper strips placed around clean towels in hotels, the removal of which helps house-keepers to tell whether a towel has been used and therefore needs replacing.

In other words, they are precautions against unintentional human error. Ideas for such measures usually arise in the minds of those, such as machine operators, working in close proximity to the asset involved. The 'Quality Circles' of a Total Productive Maintenance manufacturing environment, for example, facilitate the generation of such innovative yet simple ideas for improvement.

Chapter 5

BP Deepwater Horizon

5.1 CASE STUDY DEEPWATER HORIZON

Authoritative in-depth description of the Deepwater Horizon oil spill can be found in the Deepwater Horizon Study Group Report (DHSG, 2011), the BP Commission Report (2011), and the USDOI Report (2010).

While it can be argued that this section—which provides a summarized yet extensive description of the accident and the background to it—could have been made much shorter by referring to the abundant information in this literature, it is suggested that a primary data collection would be of lower quality because with the passage of time, memories would have faded and key personnel dispersed. Therefore, a secondary data analysis (a proven and widely used research method) has been used for structuring the problem. This also gives the possibility of triangulating sources—and the analysis can be easily checked by other researchers.

Students were briefed regarding the background to the disaster (Section 5.1.1 below), the technical cause of the failure (Section 5.1.2), and the consequences and their severity (Section 5.1.3). They were then divided into three groups, each of which was required to use reliability engineering techniques to analyze the failure and make recommendations based on the analytic tools that were used.

5.1.1 Background to the Disaster

On April 20, 2010 a massive explosion destroyed the Deepwater Horizon drilling rig in the northern Gulf of Mexico, followed by the leakage of millions of barrels of oil. The drilling rig (Figure 5.1), which was owned and operated by Transocean, had resumed the drilling of the Macondo well in the seafloor in the Gulf of Mexico, which at this point was at a depth of 8000 ft.

The drilling operation, which had been resumed on February 6, 2010, incorporated a blowout preventer (BOP) that consisted of hydraulic

Learning from Failures. DOI: http://dx.doi.org/10.1016/B978-0-12-416727-8.00005-9

■ **Figure 5.1** The Deepwater Horizon drilling rig in operation. *Rigzone (2010).*

valves designed to regulate the pressure of the oil and gases released from the well simultaneously with the drilling mud and seawater that are introduced into the well. The function of the BOP is to seal off a well that is being worked upon or drilled over, and this can happen during well interventions or drilling if there is an over-pressure that threatens the proper functioning of the rig. During the drilling stage, the pressure increased, due to two factors: the mud and seawater that were introduced into the well and the pressure that was produced by oil and gas coming from the well bottom. Figure 5.2 shows the components of the BOP.

Seawater and drilling mud were injected into the well to control the pressure. However, when the well pressure overcame the pressure of the mud and the seawater, the mud, accompanied by hydrocarbon gas, started to overflow on to the rig surface. Although it was controlled and diverted to the mud and gas separator, the gas source was unknown. Investigations revealed that hydrocarbon was being released through the poorly

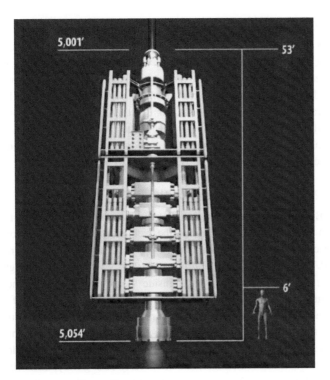

■ **Figure 5.2** Blowout preventer. *DHSG (2011).*

designed cement barrier, which was cast in order to isolate the well. The released gas mixed with mud, and at this stage the hydrocarbon, produced high pressure and damaged the gasket inside the BOP. As a consequence, a larger quantity of gas, oil, and mud flowed up to the rig surface. Minutes later, the first explosion occurred and damaged the cables that connect the control room and the BOP. As a result, the blowout was completely isolated and uncontrollable. Over time, the huge amount of hydrocarbon entered the still functioning engine room which provided the ignition source of a second, massive explosion.

5.1.2 **Technical Cause of the Failure**

Investigations indicated that there were many reasons for the disaster. However, whatever these were, the consequences were disastrous. The causes can be classified into several categories as outlined below (DHSG, 2011).

5.1.2.1 *Poor Design of Cement Barrier*

It was believed that the main cause of the explosion was the leakage of hydrocarbon from the bottom of the well to the drilling tower, which resulted in the leakage of gas into the engine room. Investigators succeeded in analyzing the scenario. From the beginning, the cement barrier that was used to prevent the gases from getting into the well was not successfully designed. In addition, it was not tested in order to see whether it could sustain the pressure. As a consequence, hydrocarbon was released through the cement barrier into the well.

5.1.2.2 *Mechanical Failure of the BOP*

The BOP is a mechanical arrangement that consists of different parts which work as valves to regulate the well pressure and prevent oil and gas from reaching the rig surface. In this case, this equipment seriously failed. The high-pressure hydrocarbon that was released through the cement layer damaged the gasket inside the BOP. As a result, the well pressure became uncontrollable. This resulted in the release of large amounts of oil and gas up to the rig surface. Mud and gas separators ventilate the hydrocarbon on to the rig surface. However, the ventilated hydrocarbon gas entered the power generator room, which provided the ignition source. Furthermore, the fire safety and gas system, which is designed to prevent fire and gas mixing, was not working or was incapable of overcoming the problem. Therefore, the explosion occurred.

5.1.2.3 *Damage of Control Cables*

There were two explosions 10 s apart. The first damaged the control cables which connected the BOP with the control room, causing the BOP to become out of control. Consequently there were no signals between the control room and the BOP.

5.1.2.4 *Fire and Gas System*

There was a safety system, to prevent fire. In oil exploration operations, frequent fire accidents are expected. However, in the Deepwater Horizon case, the system was not working, which reflected the poor maintenance of the safety systems.

5.1.3 **Consequences and Severity**

5.1.3.1 *Fatalities and Injuries*

At the time of the accident, there were 126 personnel on the Deepwater rig, and most of them did not realize that a serious problem had occurred because the safety system alarms were not functioning. They were subsequently instructed to evacuate the rig as early as possible. During that time, there was chaos on the bridge because the shift crew members were still trying to understand what was happening. In addition, emergency plan procedures were not clear to the shift crew. However, many of them were preparing to jump into the sea and others were waiting for the lifeboats to be launched. But only two lifeboats were ready and they were partially full. Unfortunately, evacuation routes were blocked because of the explosion and the atmosphere was saturated with large amounts of carbon dioxide as the ventilation systems failed. Near the rig, there was a supply ship, the Damon Bankston. It was called for help and, due to its fast response, the people who jumped into the sea were rescued first and the injured were airlifted to the nearest hospital. The consequences were 11 fatalities and 15 injuries (DHSG, 2011).

5.1.3.2 *Environmental Impact*

It is undeniable that the environmental impact of the disaster was catastrophic. The oil spill lasted for about 87 days before it was stopped. An independent analysis indicated that the total spilled oil was around 185 million gallons plus another 33.8 million gallons that were captured by BP. As a result, hundreds of miles of shoreline were oiled and large areas of the sea were closed to fishermen. Moreover, wildlife was affected by the oil spill; thousands of birds, fish, turtles, and marine mammals were killed.

5.2 **ANALYSIS OF FIRST GROUP OF STUDENTS**

The disaster happened for various reasons, some of them technical, some organizational, some financial, and yet there were problems of safety-related design.

5.2.1 **The Technical Reasons**

1. The cement barrier was incapable of preventing hydrocarbon gas from entering the well. Thus, the gas mixed with the drilling mud and was released under high pressure on to the rig surface.
2. The BOP failed to isolate the well which in turn allowed hydrocarbon and oil to seep on to the rig surface.
3. The first explosion severed the control cables between the BOP and control room. Consequently, the BOP was out of control.
4. The fire and gas system broke down, so it could not prevent the hydrocarbon ignition.

5.2.2 **The Design and Safety Reasons**

1. The well was cemented before the laboratory test results were provided to BP. Halliburton (the contractor who was responsible for testing the cement layer design) reported to BP that the well cement barrier was not effectively designed and that it was probable that a serious gas flow could occur.

2. The design of the gas and mud separator was poor. That is to say, when hydrocarbon was separated from the mud, it had not been burned or safely disposed of. It was released into the engine room where its ignition led to the massive explosion.

3. The BOP was critical equipment which was used to isolate the well and regulate the oil and gas pressure. As such it should have been redundantly designed. To be more precise, there should have been a redundant BOP connected in series. This would have reduced the probability of oil and gas leakage to the rig surface.

4. The design of the control room was poor. That is to say, influx of the hydrocarbon was not recognized until it reached the riser. The control room systems were either outdated or failed.

5. Safety system alarms were suspended (for some unknown reason), so most personnel did not realize what was happening and that the situation required evacuation.

6. The safety system had suffered various failures in the past. However, no actions had been taken to correct them.

7. There were only two evacuation boats. In the absence of evacuation procedures, most of the workers jumped into the sea.

5.2.3 **Financial and Organizational Factors**

1. In an attempt to cut costs and increase profits, BP had been cutting maintenance budgets during the previous few years. This led to inefficient maintenance programs and a backlog of deferred corrective work, especially on critical safety items.

2. Within the BP group, excessive changing of leadership positions was widespread, generating problems with work control and long-term planning.

3. Financial and time pressure was one of the main reasons for the disaster. The changing of the drilling rig took a long time and thus put workers under pressure. In addition, managers were under financial pressure as the drilling process could have taken longer than planned and the company would have been forced to pay a penalty for any overrun.

5.2.4 **Fault Tree Analysis**

Fault tree analysis (FTA) is a reliability analysis technique. Through FTA, basic events that led to the disaster can be explored. Figure 5.3 illustrates all the probable causes of the disaster. From the diagram it can be concluded that the basic causes are as follows:

1. The fire and gas system failed to detect the gas leakage, because either it failed or was overloaded.
2. The BOP failed to isolate the well for various possible reasons:
   - It was uncontrollable because the control cables were severed.
   - The operator was incompetent.
   - The design of the control panel was poor.
   - The pressure regulation valve failed to close.
3. The shoe track failed to isolate the well because it was badly designed or damaged.
4. The cement barrier was incapable of isolating the well for various possible reasons:
   - The design was not accurately tested in the laboratory.
   - The cementing process was inefficient.
   - The barrier itself was damaged by the high pressure of hydrocarbon.

Having analyzed the likely disaster causes, the next step is to evaluate the reliability of the whole system by applying the reliability block diagram (RBD) technique to an investigation of the possible design modifications that could enhance reliability.

5.2.5 **Reliability Block Diagram**

RBD is the next reliability assessment step following the FTA. It assesses the reliability of a system. Furthermore, designers can learn from the mistakes that might be the reasons for the disaster. Figure 5.4 shows, for the Deepwater Horizon case, the basic causes and their interdependencies.

5.2.6 **Recommendations and Reliability Improvement**
5.2.6.1 *Technical Recommendations*

1. Blowout preventer
   - Cover the control cables so that they can be protected against damage.
   - Install a redundant BOP, which will increase the system reliability.
   - Ensure that the BOP operator is competent.

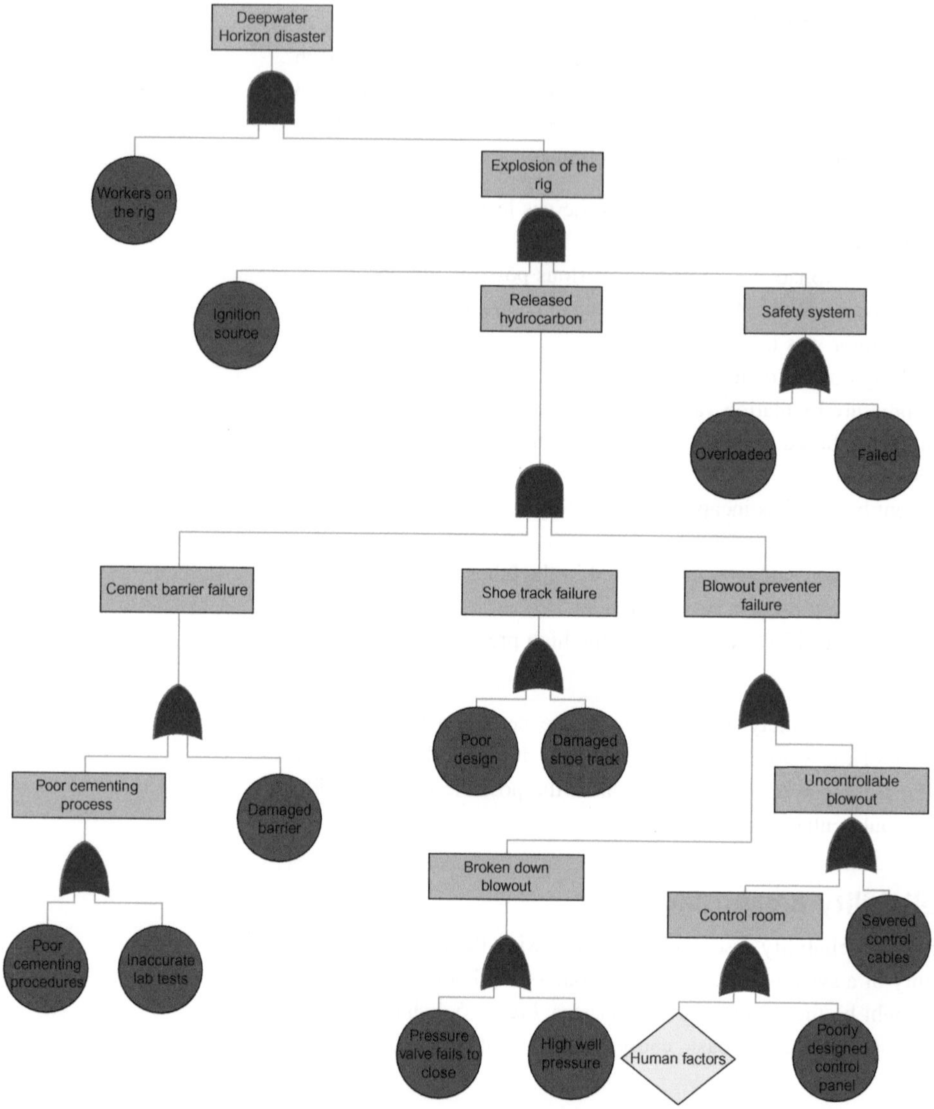

■ **Figure 5.3** The Deepwater Horizon disaster FTA.

2. Cement barrier
   - Cementing procedures need to be revised and assessed.
   - Carry out accurate experimental work to ensure the reliability of the cement barrier.
   - Study the possibility of a cement barrier replacement, e.g., using a metallic barrier.

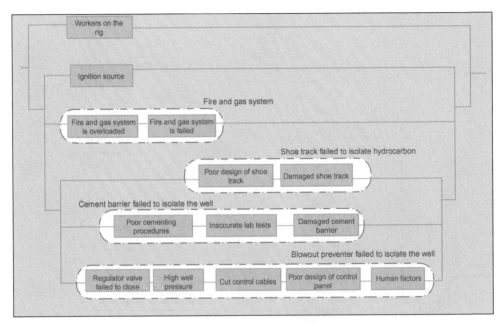

■ **Figure 5.4** RBD of Deepwater Horizon disaster.

3. Shoe track
   - Needs to be redesigned and assessed.
4. Fire and gas system
   - Replace the old system with a high capability one.
   - Install a safety system that can give an indication of a high concentration of gas.

5.2.6.2 *Management and Financial Recommendations*

- Improve skills of personnel, especially for critical tasks.
- Focus on work planning and allocate enough time for work achievement.
- Allocate maintenance budget sufficient for reliability enhancement.
- Focus on safety procedures.
- Review emergency evacuation arrangements.

5.3 ANALYSIS OF SECOND GROUP OF STUDENTS
5.3.1 Summary of the Technical Cause of the Disaster

1. Cement-casing barrier failed to isolate.
2. Incorrect interpretation of negative-pressure test.

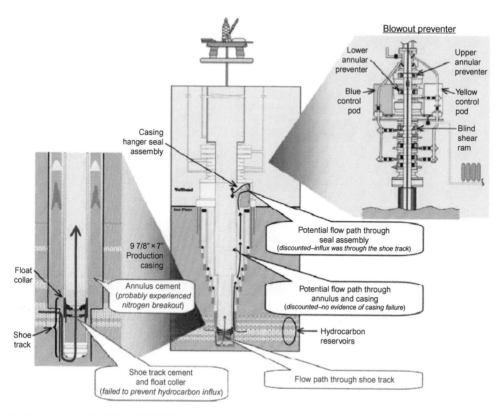

■ **Figure 5.5** Technical causes of the failure. *DHSG (2011).*

3. Fire and gas system did not prevent release of gas and its ignition.

4. BOP emergency response mode failed to seal.

These technical causes are outlined in Figure 5.5.

The resulting FTA, RBD and failure mode and effect analyses (FMEA) are illustrated in Figures 5.6−5.8.

5.3.2 **Lessons Learned**

- *Human*: Best practice and heightened vigilance in all steps from design to any variations during drilling are essential.
- *Use better technology*: Independent BOP activation control (e.g., audio-transmission), from either onshore or standby vessels, is desirable.
- *Installation*: Ensure equipment and materials (e.g., cement) are used correctly.

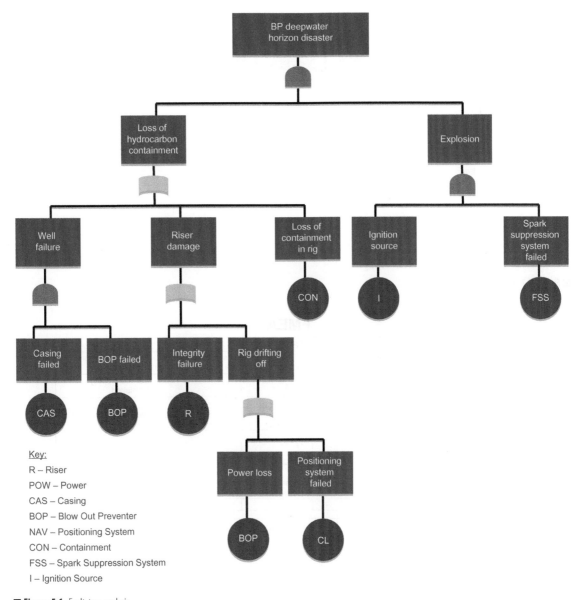

■ Figure 5.6 Fault tree analysis.

- *Redesign*: Improve BOP so that it can show its real-time conditions.
- *Data transmission:* Transmission of data onshore so that situation can be further monitored.
- *Contingency*: Backup blind shear ram (BSR), i.e., introduce redundancy.

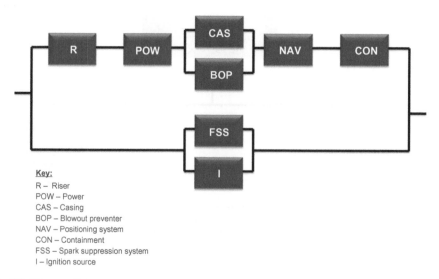

Key:
R – Riser
POW – Power
CAS – Casing
BOP – Blowout preventer
NAV – Positioning system
CON – Containment
FSS – Spark suppression system
I – Ignition source

■ **Figure 5.7** The RBD of Deepwater Horizon.

FMEA

Unit	Function	Failure mode	Failure effects	Failure causes
Blow out preventer (BOP)	To disconnect rig from well and seal well in emergency case	Well does not seal in emergency case	Oli spill to environment loss of product (pollution)	Shear Ram component undersized
				BOP not properly maintained
Casing	To provide structural support for well and sealing from hydrocarbon ingress	Well blowout	Oil and gas release to rig and environment	Imbalance of pressure in well
				Poor cement strength
Ignition Supression System	To prevent the creation of a spark	Supression system not activated	Explosion	Supression system design inadequate

Unit	Current controls	O	S	D	RPN	Recommended actions
Blow out preventer (BOP)	Redundant Ram installations	1	10	9	90	Redesign BOP shear Ram to cut through pipe and drill pipe
	Unknown	7	10	1	70	Implement regulation for independent auditing of maintenance practices
				Total	160	
Casing	1. Negative pressure test 2. Positive pressure test	4	5	2	40	Create industry standards for conducting well integrity testing
	1. Installation procedure 2. Laboratory test	2	5	2	20	Implement regulation for independent observation of cement installation
				Total	60	
Ignition Supression System	Manual operation	2	7	7	98	Redesign for automatic activation of system on sensing gas in riser prior to reaching RIG.
				Total	98	

■ **Figure 5.8** FMEA model of Deepwater Horizon.

5.4 **ANALYSIS OF THIRD GROUP OF STUDENTS**

5.4.1 **Technical Cause of the Disaster**

- Design and engineering flaws
  - BOP poor design had been indicated previously: (failed in 2006, 50% of usage).
  - Incorrect cement composition: use of nitrite and other additives reduces the strength of the required composition.
- Poor communication
  - Lack of effective communication between parties (BP, Halliburton, and Transocean).
- Poor management decisions
  - Deviation from design specification (only 6 centralizers used rather than the recommended 21).
  - No dedicated BP representative.
- Mechanical failure
  - BOP failure (of battery, BSR, and emergency disconnect system).
- Operational implementation
  - Poor judgment of pressure reading (drill line reads 1400 psi and kill line reads 0 psi) (BP and Transocean personnel could not have analyzed and interpreted pressure reading, thinking it false).
  - No action was taken due to lack of investigation of pressure reading.
  - Well integrity not established.

Figure 5.9 shows the barriers that were breached and subsequently led to the disaster. Figure 5.10 show the equivalent FTA. Figure 5.11 shows the related RBD with its estimated probabilities.

Based on Figure 5.11, one can then estimate overall (un)reliability, namely:

Reliability calculation:

> Line 1: $0.5 \times 1 - (1 - 0.4925)^2 = 0.2442$
> Line 2: $0.2425 = 0.2425$
> Line 3: $0.985 \times 0.2925 = 0.288$
> Line 4: $1 - (1 - 0.5)^3 = 0.8768$
> Line 5: $1 - (1 - 0.4851)^2 = 0.7348$
> $R = 1 - (1 - 0.2442)(1 - 0.2425)(1 - 0.288)(1 - 0.8768)(1 - 0.7348)$
> $R = 0.986$

Therefore: Unreliability

> $U = 1 - 0.986$
> $U = 0.0133.$

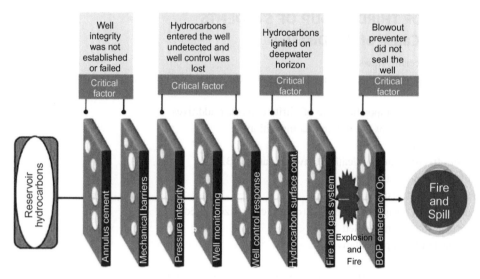

■ **Figure 5.9** Barriers for Deepwater Horizon. *DHSG (2011).*

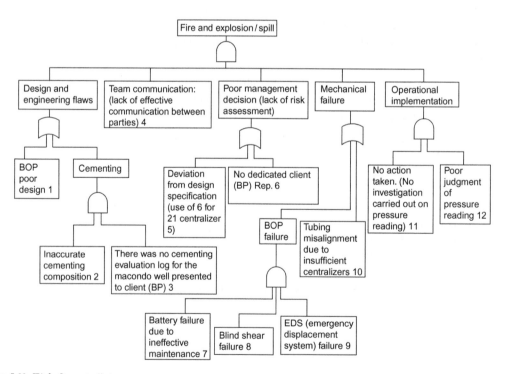

■ **Figure 5.10** FTA for Deepwater Horizon.

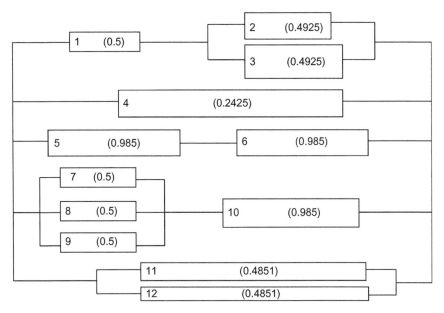

■ Figure 5.11 RBD for Deepwater Horizon.

5.4.2 **Lessons Learned**

There was evidence of the following deficiencies in the management of the operation:

- Communication gap between parties.
- Noncompliance with design specification and procedures.
- Lack of peer review.
- Ineffective process safety management.
- Lack of monitoring and supervising operational team, from client to contractors.
- Disregard for team members' judgment.
- Insufficient organizational cultural change (from focus on cost saving to focus on ensuring safety).

Improvements needed

- Full compliance to design specification and requirements (adequately centralized).
- More advanced BOP design.
- More effective communication.
- Establishment of ongoing risk assessment procedures.
- Appointment of dedicated client representative.

5.5 **FEEDBACK AND GENERIC LESSONS**

Here the tutor (the author of the current work) offered feedback to the three student groups. This was intended to provide more insight into the accident—as an incident, as a company, and as a culture—hence addressing both specific and wider generic lessons.

Each group produced a slightly different mental model, although each was exposed to the same narrative. The main differences were in the level of detail each group went into with respect to the FTA, and its use of qualitative analysis (the strength of Group 1) as opposed to its emphasis on quantitative modeling using FMEA (the strength of Group 2) and use of reliability calculations (the strength of Group 3). Nevertheless, on the whole, there were more agreements than otherwise between the findings of the groups.

The students were then directed to read a well-balanced review of quantitative versus qualitative analysis of risk (Apostolakis, 2004), and a good account of possible approaches for assessing and managing risks from extreme events (Bier et al., 1999).

It was also noted that the RBDs of all three groups had an AND-gate at the top which signifying a parallel structure. As mentioned earlier, this implies the existence of a set of barriers that were all insufficient to prevent the disaster (as illustrated in the Swiss cheese model proposed by Reason, 1997).

As regards to specific regulatory lessons, the American regulatory agency of the industry—Minerals Management Service (MMS)—had dual roles which can be seen as conflicting. On the one hand, it had a role in promoting oil and gas leasing and collecting revenues, while on the other hand it acted as regulator and enforcer on the other (Flouron, 2011). Since the accident the MMS dual functions have been separated.

In terms of BP as a company and culture, the report by the Deepwater Horizon Study Group (DHSG, 2011) concludes that this disaster also has eerie similarities to the BP Texas City refinery disaster as outlined in the BP Texas City Report (2007). These similarities include:

a. multiple system operator malfunctions during a critical period in operations;
b. not following required or accepted operations guidelines (the so-called casual compliance);
c. neglected maintenance;
d. instrumentation that either did not work properly or whose data interpretation gave *false positives*;

e. inappropriate assessment and management of operations risks;

f. multiple operations conducted at critical times with unanticipated interactions;

g. inadequate communication between members of the operations groups;

h. unawareness of risks;

i. diversion of attention at critical times;

j. a culture with incentives that provided increases in productivity without commensurate increases in protection;

k. inappropriate cost and corner cutting;

l. lack of appropriate selection and training of personnel;

m. improper management of change (as outlined by the Baker Panel, 2007; Hopkins, 2009).

The DHSG (2011) report also suggested that in both cases—Texas City and the Deepwater—meetings were held with operations personnel at the same time and place as the initial failures were developing. These meetings were intended to congratulate the operating crews and organizations for their excellent records for *worker safety*. Both of these disasters have served—as many others have served—to clearly show there are important differences between *worker safety* and *system safety*. One does not assure the other.

The DHSG (2011) report contains four strategic findings. Firstly, it warns about the apparent underestimation of the hazard caused by the "next generation" series of oil and gas exploration in the ultra-deep waters of the northern Gulf of Mexico. Secondly, it warns that the consequences of major offshore oil and gas system failures can be several orders of magnitude greater than those of previous generations of these activities, and that major developments are needed to address the consequences of major failures. Reliable systems are needed to enable effective and reliable containment and recovery of large releases of hydrocarbons in the marine environment. Thirdly, it suggests that strategies related to nonrenewables as an energy source need to be reexamined. Fourthly, it encourages collaborative efforts in benchmarking best practice in safety.

It is clear from the analysis that the Deepwater Horizon operation had many lines of defense (safety barriers) that were breached. This view is supported by research on the Texas City refinery disaster report (Holmstrom et al., 2005) which identified several deficiencies in observability and diagnosability of the condition of safety barriers in the refinery's isomerization unit.

Since safety barrier problems were common to both accidents, and both accidents are attributed to BP, students were then provided with some theory on safety barriers as regards their definition, classification, and performance (Sklet, 2006; Svenson, 1991). For example, according to Svenson, a safety barrier represents a function (and not for example an object) which can arrest the execution of the accident so that the next event in the accident chain is never realized, while a barrier system maintains the barrier function.

Also as feedback, the students were exposed to some related industrial standards that address reduction of risk through safety barriers. For example, with respect to the concepts of risk prevention, control and mitigation, ISO:13702 (1999), a standard for the oil and gas industry states that *prevention* means reduction of the likelihood of a hazardous event, *control* means limiting the extent and/or duration of a hazardous event to prevent escalation, while *mitigation* means reduction of the effects of a hazardous event. Hence, a barrier function is a one planned to prevent, control, or mitigate undesired events or accidents (Sklet, 2006).

In another development regarding barriers related to software systems, Leveson (1995) distinguished between three types of barrier function: lockout, lockin, and interlock. A *lockout* "prevents a dangerous event from occurring or prevents someone or something from entering a dangerous area or state," a *lockin* is "something that maintains a condition or preserves a system state," while an *interlock* serves "to enforce correct sequencing or to isolate two events in time."

As outlined by the Department of Environment (DOE, 1999), when analyzing barriers in an accident investigation, the following topics need to be addressed:

i. barriers that were in place, and how they performed;
ii. barriers that were in place but not used;
iii. barriers that were not in place but were required.

It is important that after their analysis of the deepwater event, students be made aware of the above theories regarding safety barriers. In this way they can better appreciate the relevance of these concepts.

5.6 **CRITICAL COMMENTARY SECTION**

It is noted that not everybody agrees with the findings and conclusions of each chapter. So at the end of some of the chapters there is an alternative view, which is not intended to undermine the issues discussed but rather present other perspectives. The ideas included do not necessarily represent the author's own view, but they are worth consideration and worth initiating some reflection and debate.

CRITICAL COMMENTARY : REDUNDANCY LEVELS

Unfortunately, many failures happen because either we have too much redundancy or give little attention to the redundancy built in at the final, downstream, stages of a process. So failures happen because of the existence of too many safe systems, an ironic situation. Levels of redundancy downstream are those which are considered as 'backups of the backups', the last resort, and the arrangement should therefore be extremely reliable - although the probability of using it is usually extremely low because of the existence of prior lines of defence. Thus, as far as possible, it should not rely on a switching mechanism, because this could undermine its reliability, ie a passive safety system. For example, a safety system should not, ideally, be activated by a certain pump or instrumentation. A better approach would be to use the failure mode itself as a source of activation or even to utilise a natural phenomenon, such as gravity, to activate the fail safe system. This minimizes the reliance on a 'switch' which, whether human, software or hardware activated, could itself be prone to failure. In the Deep Water Horizon accident the fail-safe Blowout Preventer did not activate to seal the well because the switch, located in the platform control room, failed due to fire.

BP Texas City Disaster

6.1 WHAT HAPPENED

On March 23, 2005, disaster struck at BP's Texas City oil refinery when
a series of explosions and fires resulted in the deaths of 15 people and
injured more than 170. The explosions are believed to have occurred during the start-up of an isomerization (ISOM) process unit (BP, 2005).
BP's Texas City plant was its largest and most complex refinery, with
production of up to 11 million gallons of gasoline per day and had a rated
capacity of 460,000 barrels per day (bpd). This was in addition to the
production of jet fuels and diesel fuels.

The refinery's 1200 acre site was home to 13 different process units with
around 1800 permanent BP staff. It was the United States' third largest
refinery, being owned and operated until 1999 by Amoco and still using
that company's safety management systems (which were employed prior
to the 1999 merger of Amoco and BP).

As mentioned earlier, the incident began in the ISOM unit and involved its
Raffinate Splitter, Blowdown Drum and Stack (BP, 2005). There had been
a number of previous events involving ISOM units: mainly hydrocarbons
leaks, vapor releases and fires. What principally draws attention is the fact
that two major incidents took place a few weeks prior to the fatal event on
March 23 (Khan and Amyotte, 2007). The first, which occurred in
February 2005, was a leak of hydrocarbons into the site sewer during de-
inventory of the splitter; the second occurred in the same month as the fatal
event and was a serious fire. Before going on to investigate the causes of
the incident, and the sequence of events which led to it, it is important to
form an understanding of the nature of the process which resulted in the
primary ignition of the explosion and fire on March 23.

6.2 THE PROCESS

A brief description of the processing plant and of the process itself is presented here to aid in understanding the incident. A key unit is the

Learning from Failures. DOI: http://dx.doi.org/10.1016/B978-0-12-416727-8.00006-0

Raffinate Splitter; this consists of a single fractionating column with 70 distillation trays and a feed surge drum, fire heated reboiler, and overhead condenser (Mogford, 2005). The column processes up to 45,000 bpd of raffinate from an aromatic recovery unit. The blowdown system consists of relief pipework headers, pumps, and a Blowdown Drum and Stack, the purpose of which is to receive, quench, and dispose of hot liquid and hydrocarbon vapors from the ISOM vents and pump-out systems, during process deviations or shutdowns (Khan and Amyotte, 2007). The ISOM unit works by heating the feed hydrocarbons (pentane and hexane) in order to convert them into isopentane and isohexane (to boost the octane rating of gasoline). The ISOM unit performs the conversion as the raffinate section prepares the hydrocarbon feed into the ISOM reactor (Holmstrom et al., 2006).

6.3 SEQUENCE OF EVENTS AND INCIDENT

Everything began when the operators overfilled the distillation column. The liquid was supposed to remain at a relatively low level in the column. However, the operators overfilled it nearly to its top. As a result, a mixture of liquid and gas flowed out of the gas line at the top of the column passed through the emergency overflow piping and was discharged from a tall vent hundreds of feet away from the distillation column itself (Hopkins, 2008). The disaster could have been avoided had a continuously burning flame been present at the top of the vent—which, ideally, is what should have happened. Any gases or mixtures erupting from the vent would then have been burned in the atmosphere and would not have formed the hydrocarbon vapor cloud which eventually ignited. However, such a flare system was not available (Hopkins, 2008).

On February 21, the Raffinate Splitter was shut down for a temporary outage and on February 26–28 was steamed out to remove hydrocarbons. On March 21, the splitter was brought back into service and was pressurized with nitrogen for tightness testing. On the next night, cold feed was introduced to establish feed drum and column levels.

As work started on March 23, after normal operation tasks, temperature and pressure levels in the splitter were seen to be higher than normal, so the operators opened an 8-in vent valve and witnessed steam-like vapors coming out. After doing so, pressure and temperature dropped to normal levels (Khan and Amyotte, 2007). As flow of raffinate product commenced the shift supervisor suggested opening the 1.5-in vent valve to vent off nitrogen. After shutting the vent again, the pressure peaked to 63 psig and as a result, the overhead relief valves opened to feed into the unit, which resulted in vapor and liquid coming out at the top of the stack (Khan and Amyotte, 2007). Hydrocarbon liquid and vapor were discharged to the Blowdown Drum, which eventually overfilled, leading to a geyser-like release from the stack. The final result was a series of multiple explosions and fires as the liquid hydrocarbons pooled on the ground and released flammable vapors ready to be ignited (Holmstrom et al., 2006).

As mentioned earlier, there was no flare system present, this is because the blowdown system itself was an out-of-date design installed in the 1950s and was never connected to a flare system to safely burn any released vapors. The source of the ignition of the accumulated vapor cloud is believed to have been a nearby truck engine left idling. The plant and processes involved in the incident are shown in Figures 6.1 and 6.2. The instrumentation is featured in Table 6.1.

On March 22, 2005, the Raffinate Splitter was put into operation after a month of maintenance work. Around 2.15 a.m. on March 23, cold liquid began to be fed into the Splitter at a charging rate of 15,000 bpd. By 2.38 a.m. the Splitter's sensor indicated that the liquid level had gradually increased. At that moment, the operator charged the liquid into the Raffinate Splitter to more than the 9 ft level. The normal procedures only required the tower to be filled with liquid to 6.5 ft.

Because the Splitter level indicator could measure only up to 9 ft, any level above 9 ft would not be measured by the system. The first alarm, Level Alarm Low (LAL), did trigger when the liquid passed the 6.5 ft mark but the second alarm, Level Alarm High (LAH) failed to trigger when the liquid passed the 8 ft mark. After a certain period of time, the operator stopped charging the liquid into the Raffinate Splitter. But it is estimated that at that time the liquid level went up to 100% beyond the normal operating level. The night shift operator then packed the liquid in the Raffinate Splitter by shutting off the overflow valve. As the night shift ended, the operator left the Satellite Control Board without reporting to the staff that the LAH was faulty.

When the day shift operator took over the duty, he found that there was no indication about the liquid level in the Raffinate Splitter. As they prepared to begin the start-up process, he introduced more liquid into the Raffinate Splitter, which had already been overfilled during the night shift. The liquid was then heated by opening the Reboiler flow control valve to charge the liquid into the Reboiler circuit as part of the normal start-up process. As the start-up process continued, the Splitter steadily filled with liquid up to 98 ft, but the improperly calibrated Level Transmitter (LT) indicated that the liquid level was gradually falling.

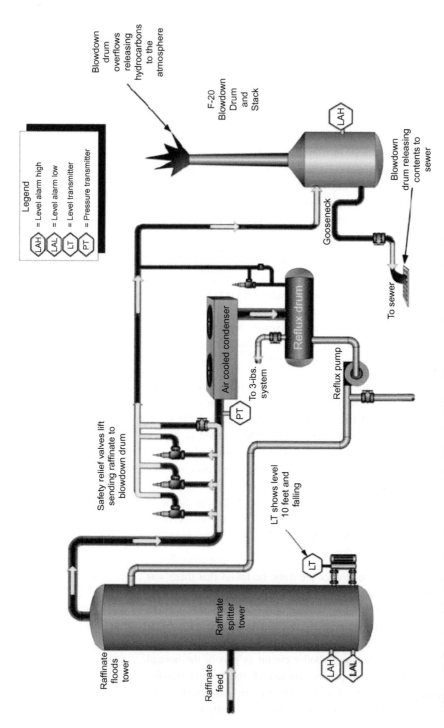

■ **Figure 6.1** Illustration of the BP Texas City Refinery Incident. *Adapted from the CSB Report (Holmstrom et al., 2006) and Mogford (2005).*

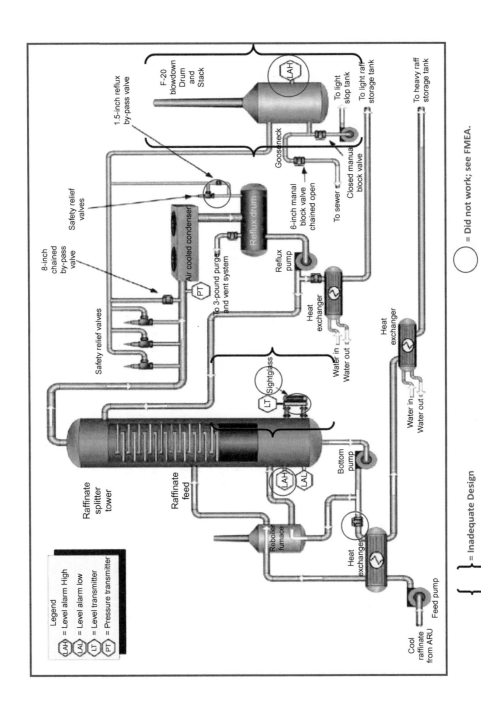

Legend

LAH = Level alarm High
LAL = Level alarm low
LT = Level transmitter
PT = Pressure transmitter

Raffinate splitter tower

Raffinate feed

Reboiler furnace

Bottom pump

LAH LAL

LT Sightglass

Heat exchanger

Cool raffinate from ARU

Feed pump

Heat exchanger

Water in
Water out

Water in
Water out

Heat exchanger

To heavy raff storage tank

Air cooled condenser

PT

8-inch chained by-pass valve

Safety relief valves

To 3-pound purge and vent system

Safety relief valves

1.5-inch reflux by-pass valve

Reflux drum

Reflux pump

Heat exchanger

Water in
Water out

F-20 blowdown Drum and Stack

LAH

Gooseneck

6-inch manal block valve chained open

To sewer

Closed manual block valve

To light slop tank

To light raff storage tank

= Inadequate Design

◯ = Did not work; see FMEA.

■ **Figure 6.2** Illustration of the BP Texas City Refinery Incident showing deficiencies in design and instrumentation. *Adapted from the CSB Report (Holmstrom et al., 2006) and Mogford (2005).*

Table 6.1 Failure Mode and Effect Analysis (FMEA) of Instrumentation That Failed

Instrumentation that failed to operate properly on March 23, 2005

Tag No.	Instrument	Function	Failure Mode	Likely Failure Cause	Effect
LT-5100	Raffinate splitter level transmitter	Transmits a signal to the control system to indicate the level in the tower	Incorrect reading prior to the incident	Instrument not calibrated for actual specific gravity of the ISOM process fluid, at operating temperatures	Transmitter falsely showed the level in the tower bottoms below 100% and falling, when in fact the tower was overfilling
LSH-5102	Raffinate splitter redundant (hard-wired) high level alarm	Alarms when the level in the tower exceeds a maximum set value	Failed to signal when the tower bottom level reached the assigned set point	Worn, misaligned components bound the mechanism	Operators received no independent warning that the maximum bottom level had been exceeded
LG-1002 A/B	Raffinate splitter sight glass	Visually indicates tower level (level indication split across two sight glasses)	Sight glasses were dirty on the inside; tower level could not be visually determined	Sight glass not cleaned	Level transmitter calibration could not be effectively performed without sight glass verification. Operators had no backup to determine tower level
LSH-5020	ISOM Blowdown Drum High Level Alarm	Alarms when the level in the Blowdown Drum exceeds a set value. This was the only high level alarm for the Blowdown Drum	Failed to signal when the tower bottom level reached the assigned set point	Damaged level displacer ("float")	No warning that the Blowdown Drum level was above maximum
PCV-5002	3-lb pressure vent valve for the Raffinate splitter reflux drum	Available for operators to manually vent gases from the splitter overhead system	Valve failed to open during start-up testing	Possible actuator stem binding, or intermittent pneumatic failure	Unit was started up with a known malfunction of this pressure control valve

Source: Adapted from Mogford (2005).

As the liquid level rose, the pressure in the Raffinate Splitter started to increase, causing the High Pressure Alarm (HPA) to trigger. Unable to identify the cause of the problem, the operator then opened the Safety Relief Valves (SRVs) to reduce the pressure. The operator also turned off the Reboiler to lower the temperature inside the Raffinate Splitter.

However, the system continued to feed the liquid into the Splitter. Due to the high temperature, the liquid inside the Raffinate Splitter started to boil and swell greatly increasing the pressure on the SRV. As a result, the liquid was forced to spill out through the SRV into the Blowdown Drum.

Because the LAH at the Blowdown Drum failed to go off, the hot liquid continuously filled the Blowdown Drum. The resulting excessive amount of liquid in the Drum could not contained and erupted into the atmosphere through the Stack. Within 90 s, a large cloud of flammable vapor spread around the area and exploded, the source of its ignition being the idling engine of a pick-up truck parked nearby.

6.4 **INVESTIGATION**

According to BP's final investigation report (BP, 2005), there were four critical factors without which the incident would not have occurred or, at least, its consequences would have been minimized, namely:

Loss of containment: This involves the action taken, or not taken, which led to overfilling of the Raffinate Splitter and eventual over-pressurization and pressure relief. The end result was liquid hydrocarbon overflowing the stack and causing the vapor cloud. Elements which contributed to the overflow and the eventual pressure relief were the very *high liquid level* and *high base temperature*, in addition to the *late start-up of the heavy raffinate rundown*. The high liquid level and normal tower pressure ended up surpassing the relief valve pressures. It is believed that stopping the feed, increasing the off-take, or reducing the heat input during the earlier stages of the exercise, or a combination of all three would have prevented the incident (Mogford, 2005).

Improper Raffinate Splitter start-up procedures and lack of application of knowledge and skills: Not following specified procedures resulted in loss of process control; appropriate knowledge was not applied and there was no supervision during start-up. Deviation from operating procedures resulted in overfilling of the Raffinate Splitter. There was a 3-h delay in starting the Heavy

Raffinate Splitter and the heat was added at a higher rate than specified and before initiating heavy raffinate rundown. The maximum temperature used also exceeded that specified by the procedure. Any actions that were taken to resolve the situation were either late or inadequate and mostly even made the situation worse. Feed was never cut and no supervision was present prior to, or during, start-up. If procedure had been followed, the incident would not have happened, although there was a clear fault in the chain of command and an ambiguity in roles and responsibilities.

Inadequate control of work and of trailer positioning: During start-up, personnel working on other tasks were in close proximity to the hazard. They were not warned or evacuated and were assembled in and around temporary trailers. The Blowdown Drum and Stack had not been considered as a potential hazardous source in any of the site studies; hence trailers were located too close to the source. Personnel were not notified of the start-up process, or when the discharge occurred. Evacuation was ineffective because alarms did not sound early enough. The nearby location of the trailers, and poor communication, increased the severity of the incident (Mogford, 2005).

Inadequate design and engineering of Blowdown Drum and Stack: The use of the Blowdown Drum and Stack as a relief and venting system for the splitter near uncontrolled areas did not take into account the several design and operational changes that had occurred over time. The splitter was not tied to a relief flare system although there had been a number of opportunities to do so. Several changes were introduced to the blowdown system which reduced its effectiveness. The use of a flare system would have reduced the impact of the accident (Mogford, 2005).

The causes of these critical factors vary from the immediate to the complex (i.e., managerial and culture). Here, however, the focus will be on the direct factors discussed already. To examine how these factors combined, two different reliability and fault finding techniques will be applied.

6.5 FAULT TREE ANALYSIS AND RELIABILITY BLOCK DIAGRAM FOR THE TEXAS CITY DISASTER

The FTA of the disaster is shown in Figure 6.3. It can be seen that the four main factors that led to the disaster were:

a. Loss of containment.
b. Improper start-up procedure.

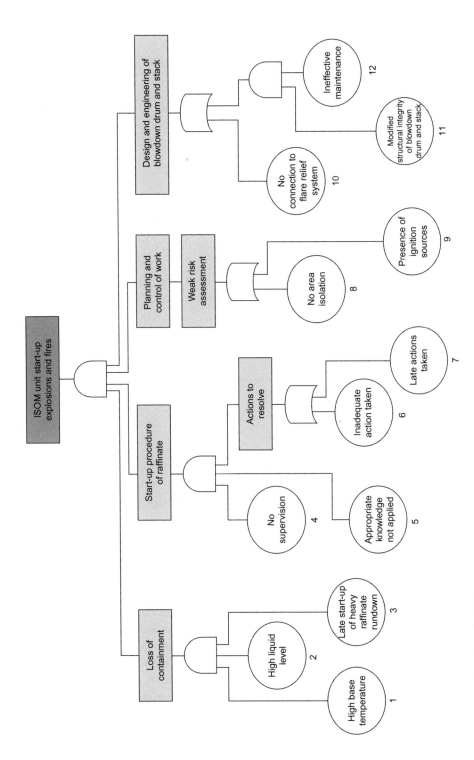

■ **Figure 6.3** Texas City incident FTA.

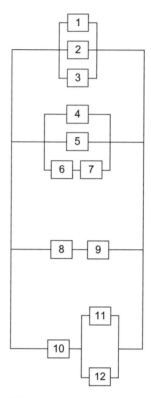

■ **Figure 6.4** Texas City incident RBD.

c. Inadequate planning and control of work, including weak risk assessment.

d. Inadequate design and engineering of blowdown drum and stack.

All these contributed to the accident and hence are linked with an AND-gate at the top of the Figure 6.3 fault tree. The narrative of the event makes it clear that the loss of containment was attributable to very *high liquid level* and *high base temperature* and the *late start-up of the heavy raffinate rundown*, and hence all three factors are linked with an AND-gate. Also, the improper Raffinate Splitter start-up was attributed to *lack of supervision*, nonapplication of appropriate knowledge, and late or inadequate and actions taken to resolve the situation, and hence the actions are further extended using an OR-gate. *Poor planning and control of work* was primarily attributed to *weak risk assessment*, which in turn caused either *a weak evacuation process or* inadequate area isolation, or improper presence of ignition sources, and hence an OR-gate models those two factors. Finally, the inadequate design and *engineering of the Blowdown Drum and Stack* was a major factor. An example of bad design was either the *splitter not being tied to a relief flare system* or the *lack of structural integrity and lack of adequate maintenance procedures*, and hence an OR-gate is used. The use of a flare system would have reduced the impact of the accident, but it was not employed even though there were a number of chances to do so. *Several changes were introduced to the blowdown system* which reduced its effectiveness and this was coupled with the *presence of ineffective maintenance procedures* and hence the use of an AND-gate.

Starting from the top of the FTA each AND-gate is modeled as a parallel structure, and each OR-gate as a series structure. It is clear from the RBD in Figure 6.4 that all four main causation factors that led to the disaster were present at the same time and hence the parallel structure (AND-gate)—which might either be a mere coincidence or an indication of many things that went simultaneously wrong, a situation that can be explained as "an accident waiting to happen." It is also clear from Figure 6.4 RBD model that Items 8 (*No area isolation*) and 9 (*Presence of ignition sources*)—where both are attributed to the existence of *weak risk assessment*—are structured in a series configuration, which reveals the importance of their contribution to process vulnerability.

In order to prevent future failures of this sort, there would be a need to address design integrity and develop better plant operating practices and procedures. In particular, the leadership team would need to establish clear common goals with rules and procedures, emphasizing teamwork, and a one-site mentality where the leadership team should hold superintendents and supervisory employees accountable for implementation of the rules and procedures. There would also be a need to establish a risk awareness training program, ensuring that leadership responsibilities include facilitating hazard identification. There is also a need to conduct a fundamental review of the control of work process and practices.

6.6 GENERIC LESSONS

Studies have shown that levels of safety start to decline when a continuous departure from standard procedures is found to have no negative effect, in the short term. Subsequently, a cause of concern starts to be considered as a routine issue and is taken lightly (i.e., kept "in-family") as happened in the Challenger and Columbia disasters (Petroski, 2006). A history of normalization of deviance, where the wrongdoing eventually

---

Table 6.2 Summary of the Three Dimensions of the "Learning" Framework Resulting from the Texas City Disaster

Feedback from users (maintenance) to design.	Planning and control of work through proposer risk assessment.
Incorporation of advanced tools in innovative applications, such as FTA and RBD	From FTA: The four main factors that led to the disaster were:
	a. Loss of containment
	b. Improper start-up procedure
	c. Inadequate planning and control of work, including weak risk assessment
	d. Inadequate design and engineering of Blowdown Drum and Stack
	From RBD: Items 8 (No area isolation) and 9 (Presence of ignition sources), where both are attributed to the existence of weak risk assessment, are structures in a series configuration, which shows their vulnerability
Generic lessons and fostering of interdisciplinary approaches.	Leadership
	Risk awareness
	Control of work
	Skill turnover
	Workplace conditions
	Rotation of management

---

becomes treated as the normal, is a recipe for a disaster. This certainly applies to the Texas City incident, where there had been previous successful start-ups, despite not following the procedures, and which resulted in safety aspects being taken too lightly. Consequently, an incident that could have been avoided occurred (Table 6.2).

6.7 **CRITICAL COMMENTARY SECTION**

It is noted that not everybody agrees with the findings and conclusions of each chapter. So at the end of some of the chapters there is an alternative view, which is not intended to undermine the issues discussed but rather present other perspectives. The ideas included do not necessarily represent the author's own view, but they are worth consideration and worth initiating some reflection and debate.

CRITICAL COMMENTARY: CAN HUMAN ERROR BE CONSIDERED AS A ROOT CAUSE OF ERROR?

If the answer to this question is 'yes', then there are those who hold this to be a misconception. Why? A good analogy was provided by the late Professor Trevor Kletz in an interview featured on a 2008 CSB safety video 'Anatomy of a Disaster', which tells the story of the 2005 BP Texas City Refinery accident.

For a long time, people were saying that most accidents were due to human error, and this is true in a sense, but it's not very helpful. It's a bit like saying that falls are due to gravity.

This succinctly addresses a core reason why accident prevention should be focussed on looking for 'real' root causes, and not on allocating individual blame. Human error has many root causes: poor morale, lack of training, weak safety culture, and so on. Generally, one way of investigating human errors is to categorize into errors of commission, or errors of omission. In other words errors due to knowledge-base or errors due to rule-base. Professor Reason's work on human reliability is a very good source of the literature on this subject.

CRITICAL COMMENTARY CAN HUMAN ERROR BE CONSIDERED AS A ROOT CAUSE OF ERROR?

Chapter 7

Chernobyl Disaster

7.1 INTRODUCTION

The Chernobyl nuclear reactor accident is of particular significance because most people have heard of it and are interested in the details of why it occurred. Its causes serve as a litany of how not to do things, the general consensus being that it resulted from a combination of flawed design, and serious operator mistakes, in a system characterized by minimal training, weak safety culture, and Cold War isolation (Hodge, 2010).

To summarize (with, for brevity, some simplification) the basic working material, the "fuel," in both a nuclear power reactor, such as that at Chernobyl, and a nuclear bomb is uranium, and the basic energy-producing process is "nuclear fission," the splitting of the uranium nuclei when bombarded by neutrons. However, it is only the nuclei of the isotope U235 that can be readily split (i.e., are "fissile") and they constitute less than 1% of the nuclei of natural uranium; the other 99-plus percent, the nuclei of U238, are only just splittable and then only by very high energy neutrons.

For a nuclear bomb to work, most (i.e., more than, say, 80%) of the uranium in its core must consist of the fissile U235 and to achieve such a concentration from the natural starting point of less than 1.0% requires a process ("Enrichment") that is technically demanding, slow, and energy intensive. In such a bomb the nuclear "chain reaction"—in which each neutron-induced fission of a uranium nucleus yields, on average, rather more than two fresh, high energy (i.e., fast), neutrons, which will have a good chance of immediately inducing further fissions—can diverge extremely rapidly, or explosively, with an immense release of energy. A general rule of thumb is that the fission of every nucleus in a gram of uranium releases the energy released by burning ten tons of oil (Hodge, 2010).

In a nuclear power reactor, however, only a small fraction (typically 3% or 4%) of the uranium nuclei in the fuel might be U235. So, to

Learning from Failures. DOI: http://dx.doi.org/10.1016/B978-0-12-416727-8.00007-2

compensate for this and hence to fortify the chain reaction and render it possible, included with the fuel is a so-called moderator (water or graphite, say). The fast neutrons emanating from fission collide with the nuclei of this moderator and thus lose most of their energy. The "slow neutrons" thus produced then have a probability of inducing fission which is increased by orders of magnitude. Rods containing strong neutron absorbers such as cadmium or boron can be used to control the chain reaction, inserted further into the reactor for negative reactivity (causing the fission rate and hence the reactor power output to fall), or taken further out for positive reactivity (causing the power output to rise). The dynamics of such control—the *rate* of rise and fall of power—is determined by, among other things, the chain reaction's fortunate dependence for its continuation on the delayed availability of those neutrons which are born not at the instant of fission but later, out of the decay of a few very unstable fragmentary nuclei produced by some fissions. This renders reactor response slower, and reactor control easier, than they otherwise would be.

The heat produced by the chain reaction is removed from the reactor by a coolant (water, more often than not) and converted, via a heat exchanger into steam to drive a turbogenerator as in a fossil-fueled power station. At Chernobyl the operators had so interfered with the control mechanisms that the chain reaction diverged uncontrollably and explosively.

Like others in the old USSR the Chernobyl reactor used slightly enriched uranium oxide fuel, graphite moderator, and ordinary (light) water coolant. This differed from reactors in much of the rest of the world, which used light water as both moderator and coolant. For this reason the design exhibited a positive-feedback behavior, termed a positive void coefficient, which meant that at low power levels an increase in steam in the coolant resulted in an increase in the reactivity, an inherent stability problem. The accident happened during a test of a safety system.

Reason (1987) provides an interesting classification of the violations (V) and errors (E) that occurred in the vital hours prior to the explosion. This is a distinction that can be useful when examining any complex failure. A violation occurs when a rule or procedure exists, but it is ignored or not executed as it should be. An error occurs when the mental or physical judgment of someone is not as it should be or is expected to be. Serious failures in complex operations are usually the result of both violations and errors. An analogy for the distinction between these two elements of failure is that of a high wire circus performer. Errors are mistakes in judgment that make it more likely that the performer will fall. Violations are failures to observe the necessary precautions, such as wearing a harness or using a safety net (Slack et al., 2009).

7.2 WHAT HAPPENED

In 1986, two explosions occurred at the newest of the four operating nuclear reactors at the Chernobyl site in the former USSR. The full extent of the long-term damage which has been estimated to have cost £3 billion has yet to be determined. It is one of history's ironies that this, the worst nuclear accident in the world, began as a test to improve safety (Snell and Howieson, 1991). Ironically, it occurred during a simulated "loss of station power" exercise performing a test of equipment which might, in such a circumstance, prevent an accident from occurring. This led to breached procedures which, with peculiarities of the control rod and reactor design, combined to cause a dramatic power surge and subsequent explosion which lifted the 1000-ton reactor upper plate.

7.3 THE TECHNICAL AND LOGIC OF THE FAILURE

The Unit 4 reactor was to be shut down for maintenance scheduled for April 25, 1986. Authorities from Moscow then decided to use this period as a window of opportunity for carrying out an experiment within the plant. The staff on duty were ordered to carry out the experiment in order to assess whether, in the event of a loss of power in the station, the slowing turbine could provide enough electrical power to operate the main core—cooling water—circulating pumps, until the diesel emergency power supply became operative. The basic aim of this was to ascertain that the cooling of the core would continue even in the event of a power failure. This was ironic as the experiment was supposed to test the same failure mode that happened during the execution of the experiment itself. *"Unfortunately, there was no proper exchange of information and coordination between the staff in charge of the test and the safety personnel in charge of the reactors, as the test would also involve the nonnuclear part of the power plant. Hence, there were no adequate safety precautions included in their test program because the operating staff were left unalert to the nuclear safety implications of their electrical testing and its potential danger"* (World Nuclear Association, 2009).

The test program would require shutting off the reactor's emergency core cooling system (ECCS), as this component was responsible for providing water to cool the core in the case of an emergency. Although subsequent

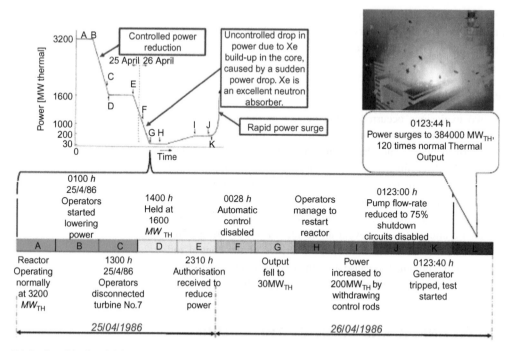

■ **Figure 7.1** Timeline of the Chernobyl disaster.

testings were not greatly affected by this, the exclusion of such a system during such a test showed the level of ignorance of the operating staffs regarding the safety of a very high class plant.

As the planned shutdown was getting in place, and the reactor was operating at about half power, the electrical load dispatcher refused to allow further shutdown, because the power was needed for the area grid. In accordance with the scheduled test program just about an hour later, the reactor was operating at half power even as the ECCS was switched off. Though it was not until about 23:00 on 25 April that the grid controller had to make a further reduction in power.

The timeline of the disaster is shown in Figure 7.1 (International Nuclear Safety Advisory Group (INSAG), 1992).

7.4 **CAUSES OF THE INCIDENT**

The causes of this disaster were investigated by the International Atomic Energy Agency, who set up an investigation team, the INSAG, which in its report gave the following as the causes of the accident (INSAG, 1986).

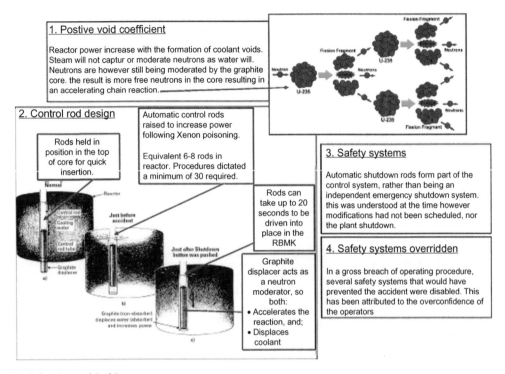

■ Figure 7.2 Technical cause of the failure.

- There were gross violations of operating rules and regulations; as within the course of preparation and testing of the turbine generator under rundown conditions using the auxiliary load, the operating staff disconnected a series of technical protection systems and breached the most important operational safety condition for conducting a technical exercise.
- Also, the reactor operators disabled safety systems down to the generator, which the whole testing program was concerned.
- Their main process computer, SKALA, was running in such a way that the main control computer could not shut down the reactor or even produce power.
- All control was transferred from the above process computer to the human operators (IAEA, 1991, 2005).

Figure 7.2 provides more details about the technical cause of the failure in terms of the positive void coefficient, control rod design, safety systems in place, and safety systems that were overridden (INSAG, 1992).

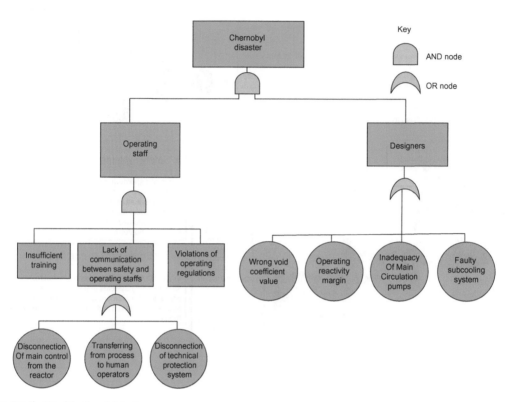

■ **Figure 7.3** The FTA of the Chernobyl disaster.

7.5 **FAULT TREE ANALYSIS AND RELIABILITY BLOCK DIAGRAM FOR THE DISASTER**

The fault tree analysis (FTA) shown in Figure 7.3 shows that the two main causes of the disaster are attributable to errors by the operating staff and designers. According to the narrative of the case study, factors contributing to errors by the operating staff are lack of training, lack of communication between safety and operating staff, and violation of operating regulations; all linked with an AND gate as they occurred simultaneously. The lack of communication is evidenced by the disconnection of the main control from the reactor, the transfer from process to human operators, and the disconnection of the technical protection system without the involvement of the operators. As for the issue of bad design, possible causes of the accident are attributable to wrong void coefficient value, insufficient operating reactivity margin, inadequacy of main circulation pumps, or faulty subcooling system, joined by an OR gate.

AND = Parallel
OR = Series

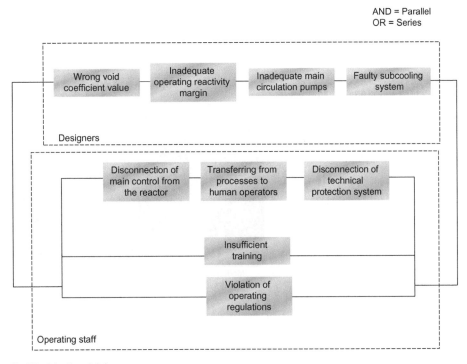

■ Figure 7.4 The RBD of the Chernobyl disaster.

The reliability block diagram (RBD) shown in Figure 7.4 shows that within the operating staff, there was a particular problem in the lack of communication between safety and operating staff, the more RBD boxes in series the less reliable the system. The same applies to the issue of bad design, evidenced by the four boxes connected in series.

7.6 GENERIC LESSONS AND PROPOSED IMPROVEMENTS

- Designers need to simulate their designs and carry out exhaustive testing.
- Operating staff must be adequately and intensively trained in safety, operations, and troubleshooting.
- Good communication between operations and safety personnel is essential during nonnormal operations (such as the testing program in this case).

Modern safety improvements include:

1. Independent control and shutdown systems.

2. Simple yet powerful mechanical design— i.e., gravity or spring fired shutdown rods.

3. Multiple and diverse shutdown systems:
   - Control rods
   - Boronated water/beads.
   - Neutron poison injection systems.
- Attempts to bypass these systems results in reactor trip.

4. Shutdown system must be tested on-power and meet stringent reliability targets.

5. Safety systems cannot be overridden.

Gas turbine /diesel back-up generators supply emergency power for cooling at modern stations, taking seconds to initiate. This removes the need for the running down turbine to power cooling pumps in the meantime as at Chernobyl.

Modern designs includs full core containment capable of withstanding several atmospheres of pressure.

More and more safety features are designed to be passive i.e., Naturally occurring (no operator action or electronic signal required) following a given occurrence. For example, positive void coefficient reactors (such as the RBMK) are banned in North America, a loss of primary coolant in American PWR's removes moderation and as such halts any ongoing chain reaction.

■ **Figure 7.5** Summary of lessons learned.

- Consultation between designers, original equipment manufacturer (OEM), and the operators should be held before conducting test operations such as described here.
- Operating staff should be required to follow the design procedures.

Figure 7.5 summarizes the lessons learned.

The cause of Chernobyl finds its roots in organizational factors. According to the IAEA, "poor safety culture" was a primary cause (Cox and Flin, 1998). Also, according to Ahearne (1987), Chernobyl can be informatively compared with the 1979 accident at the Three Mile Island nuclear power station. He found significant similarities between the two accidents including complacency of operators and the nuclear industry, deliberate negation of safety systems, and operator's lack of understanding of their plant, all of which shows the critical importance of the human element.

Table 7.1 The Three Dimensions of the "Learning" Framework as an Outcome of This Disaster Case Study	
Feedback from users (maintenance) to design.	Stronger communication between designers, engineers and operational personnel.
Incorporation of advanced tools, such as FTA and RBD	a. From FTA: The two main causes of the disaster were attributed to errors by the operating staff and designers.
	b. From RBD: There was particular problem of communication between safety and operating staff.
Generic lessons and fostering of interdisciplinary approaches.	Design simulation.
	Adequate training.
	Improved communication.
	Consultation between stakeholders.
	Adherence to procedures.

Although Chernobyl disaster was a particularly dramatic failure, it is possible to draw out some more general issues that apply to less disastrous incidents. In particular, this can contribute to our understanding of:

- The specified routines, rules, and procedures that are intended to ensure that failure does not occur.
- The importance of the implicit knowledge and learning that are required if the chance of serious failure is to be minimized.
- The skills necessary both to prevent failure and to recover from it.

Regarding design, lessons learned, which are also now incorporated in new designs, are (Snell and Howieson, 1991):

- Keep the control and shutdown systems independent. The Chernobyl reactor was difficult to control and relied too much on operators for that control.
- Keep the mechanical design simple and robust.
- Ensure that shutdown systems can be tested on-power, to ensure that they meet stringent reliability targets.

An interesting way of summarizing the Chernobyl event has been offered by Reason (1987): *"Together they made a dangerous mixture: a group of single-minded, but not nuclear, engineers directing a team of dedicated but overconfident operators. Each group probably assumed that the other knew what it was doing. And both parties had little or no understanding of the dangers they were counting, or of the system they were abusing."* These comments are truly generic in that they can well be applied to many other contemporary industrial situations (Table 7.1).

7.7 **CRITICAL COMMENTARY SECTION**

It is noted that not everybody agrees with the findings and conclusions of each chapter. So at the end of some of the chapters there is an alternative view, which is not intended to undermine the issues discussed but rather present other perspectives. The ideas included do not necessarily represent the author's own view, but they are worth consideration and worth initiating some reflection and debate.

CRITICAL COMMENTARY: FIT A PROBLEM TO THE MODEL

Some may argue that the danger of using specific analytical tools and models to study a disaster limits the knowledge gain, a complex situation being unnecessarily simplified, and to a degree that may limit our ability to comprehend the complexities of the interactions between the various influencing factors. Others, on the other hand, may argue that a hybrid use of models can overcome the limitations of a 'mono-model', such limitations cancelling each other. Hybrid modelling can thus provide a way to view the problem from multiple perspectives and hence facilitate a more comprehensive approach. Although these views would appear to be contradictory, it could be argued that, to some extent, both are valid.

Chapter 8

The Concorde Crash

NOTATION

γ FMEA rating
α Modified risk priority number (RPN)
α_{av} Average modified risk priority number

8.1 INTRODUCTION

Normally, it has been difficult to combine both quantitative and qualitative judgments when making large complex decisions. In the case presented here, both sets of data, quantitative and qualitative, are combined in the analytic hierarchy process (AHP) to arrive at an overall decision. Subjective inputs are entered directly via pairwise comparisons, while qualitative inputs are first acquired through the use of a modified FMEA known here as "γ analysis."

Traditional FMEA (O'Connor, 1990; Stamatis, 1995) has been criticized for having several drawbacks; it concentrates on the analysis of existing systems rather than proposing ways of designing-in system excellence (Aven, 1992; Price and Taylor, 2002; Pillay and Wang, 2003). Other conventional reliability analyses, such as fault tree analysis and overall equipment effectiveness, are based on the analysis of plant details in a rigid and static environment. They have been successful to a degree, but the volume and quality of data involved render them expensive and ineffective solutions to design problems.

The model proposed here acts as an alternative, offering a flexible and intelligent approach using criticality analysis on a multilevel and multi-axis view to focus on appropriate improvements and to select appropriate design policies. In previous work (Labib, 1998; Labib et al., 1998a,b). Multiple criteria decision making has been applied to maintenance problems. Application of FMEA for feedback into design has been addressed by others (Franceschini and Galetto, 2001; Farquharson et al., 2002).

Learning from Failures. DOI: http://dx.doi.org/10.1016/B978-0-12-416727-8.00008-4

8.2 **THE ACCIDENT**

The first and only recorded loss of a Concorde occurred on July 25, 2000. Flight AFR 4590 was involved in a catastrophic accident soon after takeoff from Charles de Gaulle airport, France, causing complete destruction of the airframe and the loss of 113 lives (9 crew members, 100 passengers, and 4 on the ground). All remaining aircraft were immediately grounded to allow execution of a thorough investigation into the accident. This not only led to a number of hypotheses about the causes of the accident being presented and tested, but also recommended that the aircraft could be allowed to fly again if suitable modifications were made.

It was concluded that the accident was caused by the rupture and subsequent disintegration of the front right tire of the left main landing gear after it was punctured by a metal strip which had fallen off the thrust reverser cowl door on a DC10 that had taken off 5 min before Concorde. It was deduced that the tire failure initiated a sequence of events which led to leaking fuel catching fire, severely damaging the airframe in flight and culminating in a loss of control before the aircraft made contact with the ground.

The following is a brief synopsis of the events:

- A tire was damaged during takeoff throwing rubber fragments against the aircraft wing at high velocity.
- Fuel Tank Number 5 was ruptured and spewed out fuel which then caught fire.
- The landing gear would not retract, and ingestion of hot gases and kerosene by the engines caused power surges and a consequent loss of thrust.
- The aircraft angle of attack increased rapidly, increasing drag, and leading to lift being rapidly lost.

The fuel was likely to have been ignited from an electrical spark, an engine surge, or from self-ignition of hot gases.

The rupture of a fuel tank after being hit by *tire* rubber was an unusual problem and perhaps exclusive to Concorde due to its very high takeoff velocity, which was in excess of 200 mph. Despite this, the accident should have been prevented if adequate steps had been taken after similar earlier failures. There are several ways to protect fuel tanks, as well as trying to ensure tires do not rupture. However, how to decide which is the most effective? The model introduced here addresses this problem.

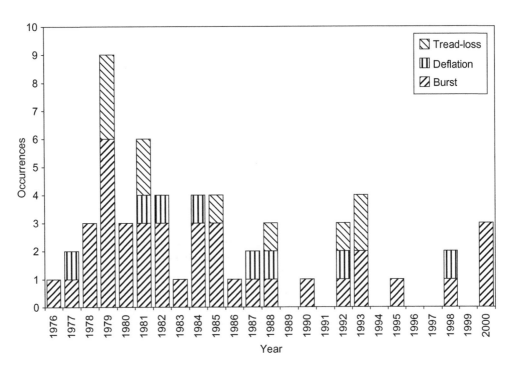

■ **Figure 8.1** History of events concerning tire failure on Concorde. *Bureau d'Enquetes (2002).*

8.2.1 **A History of Problems**

It is prudent to note that this was not the first identified case of a tire rupturing and damaging Concorde during takeoff (Figure 8.1), although this was the first where a failure had a catastrophic outcome.

Previous modifications to reduce the risk at takeoff included the ending of the use of retreaded tires, by British Airways in 1981 and then Air France in 1996. This entirely eliminated the problem of tires shedding their tread (all tread-loss events between 1981 and 1996 were attributed to Air France aircraft). Other modifications included, in 1982, strengthening the wheel and tire assemblies. This again reduced the number of events involving tire failure.

8.2.2 **Options for Improvement**

To minimize the likelihood of any repeat failure of the type described, a number of major modifications were considered. The main options were as follows:

1. *To more thoroughly protect the cabling and hydraulics* running
   through the main landing gear bays. This would allow full operation

of aircraft systems in the event of a repeat high-velocity impact directed at the gear bay. This would be done in conjunction with automatically disconnecting the 115 V electrical supply to the brake cooling fans while the landing gear was deployed, eliminating electrical sparks as a possible source of ignition for escaping fuel. This was a low-cost option as there would be little addition to the aircraft by way of weight or volume because tough, low-density materials could be used, such as fiber-reinforced composites.

2. *To make use of more recent tire technology by employing radial rather than cross-ply construction* which, in the event of a blowout should burst parallel to, rather than perpendicular to, the axis of rotation. This was a relatively low-cost solution which was likely to improve the safety status of the aircraft. However, potential customers might have seen the change as too minor to significantly improve the safety of the aircraft.

3. *To line the aircraft fuel tanks with kevlar—rubber panels*, which would resist impact penetration better than the then current materials. The solution was similar to that in military planes where fuel tanks are lined with rubber which allows a bullet or other body to pass through before the ensuing gap subsequently closes, minimizing fuel loss and, therefore, the potential for fire. This, although not a low-cost option, was perhaps the most robust in terms of meeting the criteria of improving safety. It would appear to airline customers that significant steps were being taken to improve the safety of the aeroplane.

4. *To permanently ground the aircraft* would have carried the minimum of risk. If there were no passengers on the aircraft, and it did not fly, there could be no risk. This last option will also be used in the following analysis. It would have been low cost. But these would have been a publicity trade-off because Concorde added prestige to British Airways and Air France aircraft fleets.

5. *To maintain the status quo*. Although not a real option, because the aircraft would not have been certified to fly, it will be used in the analysis as a check on the decision outcome.

8.3 THEORY AND USE OF γ ANALYSIS (MODIFIED FMEA)

There will be four main steps in the model that will now be formulated. The first is to perform the modified FMEA to gain a γ rating for each modification. The second is to construct a hierarchy, for use with the AHP, which includes γ as a criterion, the third is to make judgments to add to the AHP, and the fourth is to run the AHP mathematical program to synthesize the result of the analysis.

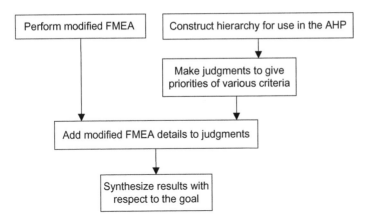

■ **Figure 8.2** Model flow chart.

The process is adaptable in its nature and each task need not be done concurrently. So, for example, the γ analysis may be completed before the construction of the hierarchy takes place or vice versa (Figure 8.2).

The principle of the FMEA remains the same when used to find the RPN for a given component or overall system, as when used normally (Dale and Shaw, 1990; Mohr, 2002) when used in the model as when used normally; to find the RPN. However, in the modified version, there are two extra stages of analysis to perform, the first being calculation of the probability of the cause of the failure occurring and the second relating to preparing the RPN data for entry into the AHP. Normally, the RPN is a value between 1 and 1000, but this is reduced to a value between 0 and 1. This is done using a logarithmic function.

It is at this stage that the new method of analysis begins to differ from a standard FMEA. The objective is to gain an overall RPN (α) for the modification option being studied, rather than an α value for each individual failure effect. This is done by averaging α for all the failure effects. As a result, all the values entered in the table (Figure 8.3) need to be scaled accordingly to allow direct comparison.

A new column is added to the FMEA table, for the probability of the cause of the failure occurring, known as the probability product, this is then assessed on a scale of $1-10$. This value is then used to scale the probability values for each failure effect by using the following formula:

Probability product = (Probability of cause/10) · probability of the effect

$$(8.1)$$

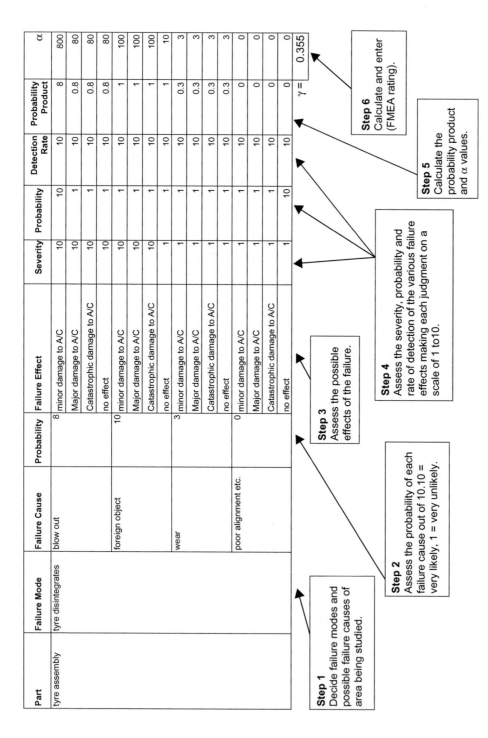

Part	Failure Mode	Failure Cause	Probability	Failure Effect	Severity	Probability	Detection Rate	Probability Product	α
tyre assembly	tyre disintegrates	blow out	8	minor damage to A/C	10	10	10	8	800
				Major damage to A/C	10	1	10	0.8	80
				Catastrophic damage to A/C	10	1	10	0.8	80
				no effect	10	1	10	0.8	80
		foreign object	10	minor damage to A/C	10	1	10	1	100
				Major damage to A/C	10	1	10	1	100
				Catastrophic damage to A/C	10	1	10	1	100
				no effect	1	1	10	1	10
		wear	3	minor damage to A/C	1	1	10	0.3	3
				Major damage to A/C	1	1	10	0.3	3
				Catastrophic damage to A/C	1	1	10	0.3	3
				no effect	1	1	10	0.3	3
		poor alignment etc.	0	minor damage to A/C	1	1	10	0	0
				Major damage to A/C	1	1	10	0	0
				Catastrophic damage to A/C	1	1	10	0	0
				no effect	1	10	10	0	0
							γ =		0.355

Step 1
Decide failure modes and possible failure causes of area being studied.

Step 2
Assess the probability of each failure cause out of 10.10 = very likely, 1 = very unlikely.

Step 3
Assess the possible effects of the failure.

Step 4
Assess the severity, probability and rate of detection of the various failure effects making each judgment on a scale of 1 to 10.

Step 5
Calculate the probability product and α values.

Step 6
Calculate and enter (FMEA rating).

■ **Figure 8.3** γ analysis table, demonstrating formulation of γ values for a tire failure.

The values derived are given in a column entitled "probability product." It is this value which is used when calculating the α for each failure effect, which is found from the expression:

$$\alpha = \text{severity} \cdot \text{detection rate} \cdot \text{probability product} \quad (8.2)$$

This allows α to be given for each failure effect, which takes account of the probability of the cause. The average of all the values is then found for a particular modification solution to give its overall rating. This α_{av} value is then used in the overall Eq. (8.3) to find γ.

As previously stated, the value of α in this form is not suitable for use in the AHP as values need to be between 0 and 1 for direct entry into the analysis. This is resolved by obtaining only one number to describe the risk associated with the part being studied. The average of the α values is found (α_{av}) which, depending on the judgments, can be any value between 0 and 1000. The following formula is then applied to reduce the range to 1−10:

$$\gamma = 1 - \left(\frac{\log_{1.9952599}(\alpha_{av} + 1)}{10} \right) \quad (8.3)$$

Notes:

- The total value of the derived expression is subtracted from 1 to give the analysis the correct sense: FMEA looks at how bad a problem is with a high value being *very bad*, whereas the AHP looks at how good a solution is with a high value being *very good*. This step corrects the above anomaly.
- The second part of the formula: $\log_{1.9952599}(\alpha_{av} + 1)$ reduces the range of values outputted from the average of α (1−1000) to a range of values from 1 to 10. 1 is added to the α_{av} to avoid the case where a log of 0 may need to be found. A base of 1.9952599 is used as this gives a range of solutions between 0 and 10 when the data entered into the formula are in the range 0−1000.
- The answer from the above step is divided by 10 to reduce the values from a range of 0−10 to a range of 0−1. This data can then be entered directly into the AHP.

For this analysis, an additional node must be added to the hierarchy in the form of γ; it is under this heading that the γ values will be added.

This criterion is assessed as before with the weighting of its importance being decided through the use of the normal pairwise comparisons. Thus, if the qualitative data is deemed to be very important, then γ will have a high weighting, if not the weighting should be lower. For systems where the γ analysis is thorough, a weighting of around 0.4 should be used, less so if the γ results are felt to be less reliable.

8.4 **APPLICATION OF THE AHP TO THE CONCORDE ACCIDENT**

The AHP has been developed by Saaty (1980, 1990, 1996) and has been applied to various problems (Dobias, 1990; Vargas, 1990; Talbert et al., 1996).

In Section 8.2.2, the following modification alternatives were listed:

- Ground the aircraft.
- Line the fuel tanks with a composite material.
- Change the construction of the tires fitted to the aircraft.
- Give additional protection to systems passing through the landing gear bay.
- Make no changes.

For the purpose of this example, the decisions are simplified by making a number of assumptions, namely that the options are mutually exclusive, and that making no changes to the aircraft is a valid option. In reality of course, this is not the case.

It may be argued at this stage that it is not appropriate to carry out an analysis on a system as complex as the modification of Concorde, especially since only those with a thorough understanding of the problem should expect to get accurate responses from using the proposed model. However, this study is intended on as a demonstration of the use of the tool; the results are not critical in themselves.

8.4.1 **Step 1: γ Analysis**

The initial step involved here was to identify the system which was to be analyzed. Therefore, the root cause of the Concorde accident, in this case a tire failure, was used to form the overall goal, i.e., to *"[ensure] appropriate measures have been taken to guarantee a satisfactory level of safety with regard to the risks associated with the destruction of tires"* (Bureau d'Enquetes, 2002).

A γ analysis table was filled in for each modification option, as described in Section 8.3. There were therefore five separate tables to be completed, which was performed in a spreadsheet.

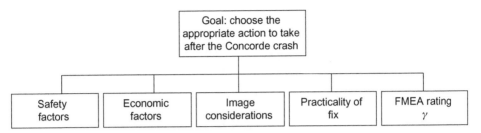

■ **Figure. 8.4** AHP hierarchy showing the goal and the child nodes of the goal.

Judgments were then made as to the probability of each cause, the severity of each effect, the detection rate of each effect, and the probability of each effect. Formulae were entered in the columns entitled "probability product," and "α" and in the cell which calculates γ, in accordance with the method described in Section 8.3. The table used to assess changing the tires, as is shown in Figure 8.3.

The assessment for "grounding the aircraft" had to be treated as a special case, because if the aircraft did not fly there could have been no risk from flying on the aeroplane, so γ would immediately have been zero. This was not reasonable; it was therefore assumed that airline customers *had* to fly and therefore the risk of flying on an aircraft other than Concorde should be subjectively assessed for this entry.

8.4.2 **Step 2: Construction of the Hierarchy**

Construction of the hierarchy (Figure 8.4) was carried out after taking account of the main factors influencing the future of the aircraft. The goal was thus deemed to be *to choose the appropriate action to take after the Concorde crash*. The goal was then assessed using the following criteria:

- *Safety*: Are the safety issues satisfied?
- *Economic*: Will the choice be economically viable?
- *Image*: Will the public be satisfied with the changes?
- *Practicality*: Are the proposals feasible?
- *FMEA*: γ.

Note that all except the γ assessment are posed in the form of a question to ease the pairwise comparisons which are made at a later stage. This is not required for γ as the data is entered directly.

The alternatives were entered in shortened version as follows:

- Line fuel tanks
- Change tires

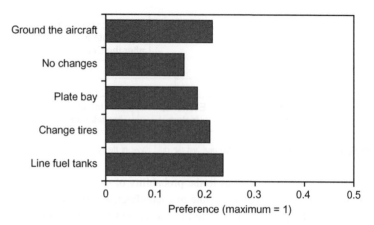

■ Figure 8.5 Results of analysis showing lining of the fuel tanks as the most preferable alternative.

- Plate bay
- No changes
- Ground the A/C

With the hierarchy completed and all alternative entered, it is possible to move to the next step of the analysis.

8.4.3 **Step 3: Making Judgments**

Each judgment was made with respect to the alternatives while being mindful of the background of the accident and the factors which would influence the decision which the engineers would make. The γ values were added to the AHP at this stage to allow the software to take account of them when synthesizing the result.

8.4.4 **Step 4: Synthesis of Results with Respect to the Goal**

The mathematical solver of the AHP was run, to synthesize the results, which are shown in Figure 8.5.

8.5 **ANALYSIS OF RESULTS**

It can easily be seen that lining the fuel tanks is the most preferable option, followed by permanently grounding the aircraft (note that both decisions were made in the same sequence). This is after all the criteria have been assessed, including the γ rating. The sensitivity of these results was then analyzed using the following graphs.

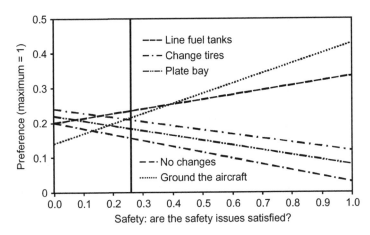

■ **Figure 8.6** Sensitivity graph: safety.

Figure 8.6 shows a sensitivity graph with the *safety* criterion on the horizontal axis. Sensitivity graphs allow the effect to be studied of changing the importance of one particular criterion while the importance of all other criteria remains unchanged.

Figure 8.6, by detailing the decision surface, demonstrates what happens when the *safety* criterion becomes more, or less important than the initial judgment. It can be seen that at the decision point (the vertical line) *lining the fuel tanks* is the best option. However, if safety was to be given less importance in relation to the other criteria (vertical line moves left) then *changing the tires* would be the best option for achieving the goal. Similarly, if *safety* issues were to become more important (vertical line moves right) then *grounding the aircraft* would be the recommendation of the analysis.

When the influence of *practicality* is considered (Figure 8.7), *lining the fuel tanks* is only considered the best option when this criterion is given a low level of importance. If it were to become more important, then *changing the tires* would quickly become the best option for fulfilling the goal. *Grounding the aircraft* is not considered a practical option as the aircraft would not be given recertification; as a result it is the least likely option for all levels of *practicality*.

Public perception of the aircraft, or *image* as it is described in this analysis (Figure 8.8), is best satisfied by *lining the fuel tanks* of the aircraft. This is due to this being a major modification to the aircraft and one which has a substantial cost, which could be used in publicity material. Making *no changes* would not be acceptable at any level and *grounding the aircraft* would also

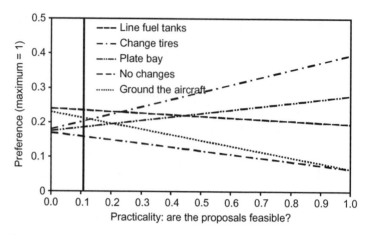

■ Figure 8.7 Sensitivity graph: practicality.

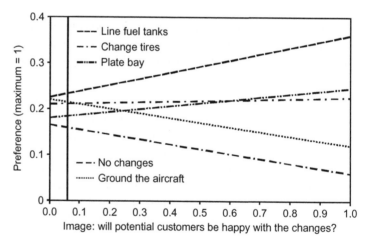

■ Figure 8.8 Sensitivity graph: image.

be reasonably unpopular because the aircraft would be perceived as a status symbol of European technology.

Economic concerns have been prescribed a low level of importance (Figure 8.9) in the initial evaluation. This is based on the historical economic nonviability of Concorde and it is seen as being a factor where the policy of the airlines is unlikely to change. In this case the *lining of the fuel tanks* would again be most popular, although if the *economic* weighting is increased, making *no changes* would very quickly become the most desirable option, for obvious reasons.

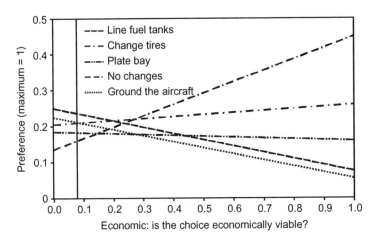

■ **Figure 8.9** Sensitivity graph: economic.

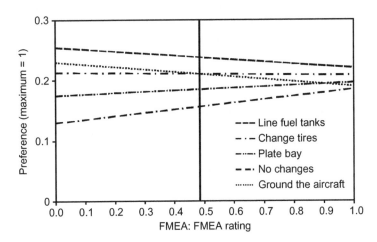

■ **Figure 8.10** Sensitivity graph: γ analysis.

Studying the gradient graph (Figure 8.10) relating to the FMEA rating (γ) shows that *lining the fuel* tanks would be the best option whatever weighting the criterion is given. At the lower end of the scale, where γ is least important, *grounding the aircraft* would become a close second, whereas at the upper end of the scale *changing the tire design* would become the second most likely choice to satisfy the goal. Studying these graphs allows the relative importance of the various criteria to be analyzed, which is a useful tool in the verification of the decision. In this case, *lining the fuel tanks* is consistently the preferred option for fulfilling the goal and as such it is reasonable to have a high level of confidence in the result of the analysis.

8.6 CONCLUSION

The method has shown itself to be thorough in its nature and, through the use of the sensitivity analysis, fully accountable. This allows any outcome to be checked and double-checked to ensure it is the most appropriate one, and a permanent record of the analysis process to be formed.

This case study has arrived at the conclusion that *lining the fuel tanks* of Concorde would have been the preferable option when deciding the future of the aircraft.

This matches the decision made by the engineers working on Concorde at the time and who decided to line the fuel tanks with a composite material. The next most preferable option was to permanently ground the aircraft, partly due to its age.

This case study has thus shown that the model presented is likely to be of use in aiding with this type of decision making and that the result is likely to be of a robust, dependable nature. The Concorde accident, although somewhat "freak" due to a combination of unlikely events, can still be analyzed in a methodical manner when using suitable models.

The use of the AHP could well prove invaluable in other cases in that it will allow a fully documented and transparent decision to be made with full accountability. It will also ease the task of justifying the decision, because it can be shown to have been made for eminently reasonable criteria.

This case study has focused on the implementation of the model and its subsequent evaluation to ensure that the decision made was robust. In this case it was possible to compare the decision made from this analysis with what has actually happened.

Modification of Concorde was completed with, lining the fuel tanks being chosen as the preferred option. This was at a cost of 17 million UK pounds. The aircraft was returned to service but was grounded permanently in 2004 due to reduced passenger revenues following a global economic downturn.

8.7 **CRITICAL COMMENTARY SECTION**

It is noted that not everybody agrees with the findings and conclusions of each chapter. So at the end of some of the chapters there is an alternative view, which is not intended to undermine the issues discussed but rather present other perspectives. The ideas included do not necessarily represent the author's own view, but they are worth consideration and worth initiating some reflection and debate.

CRITICAL COMMENTARY: ARE PROBABILITY AND SEVERITY APPROPRIATE MEASURES OF SAFETY AGAINST MAJOR DISASTER?

Although risk is defined by some as a product of measures of probability and of severity, it could be argued that, in some situations, both measures could be misleading. For example, the definition of a disaster is that it is a 'rare' event that causes great destruction. So here 'rarity' implies low probability and hence can lead to over-confidence. The appropriate measure should concern 'possibility' rather than 'probability'. Techniques such as Failure Mode and Effect Analysis need to give less weight to probability and more to the ability to detect and recover, and provide more emphasis on brainstorming to uncover possible failure mode scenarios. Also as regards severity as a measure of risk, this can again sometimes lead to a kind of a mindset that ignores 'near miss' situations because damage (severity) has yet to occur. But the problem is that how 'near' was the near miss and how lucky was the 'miss' itself? It might actually have been a case of 'an accident waiting to happen' but with our way of assessing risk just failing to capture such an event. So both probability and severity need to be treated with caution, and the traps lurking in their misuse avoided. For example, Figure 8.11 shows history of events concerning tire failures on the Concorde. In retrospective, one may argue that they were near −misses.

■ **Figure 8.11** Image of the Concorde at take-off prior to the crash.

Chapter 9

Fukushima Nuclear Disaster

9.1 BRIEF INTRODUCTION

9.1.1 The Evolutionary of the Disaster

On March 11, 2011, Japan suffered its worst recorded earthquake, known as the Tohuku event. It was classified as a seismic event magnitude 9.0, with maximum measured ground acceleration of 0.52 g (5.07 m/s^2). The epicenter was 110 miles East-northeast (ENE) from the Fukushima-1 site. Reactor Units 1, 2, and 3 on this site were operating at power before the event, and on detection of the earthquake they shut down safely.

Initially, on-site power was used to provide essential post-trip cooling. About an hour after shutdown, a massive tsunami, generated by the earthquake, swamped the site and took out the AC electrical power capability. Sometime later, alternative backup cooling was also lost.

With the loss of these cooling systems, Reactor Units 1 to 3 overheated, as did a spent fuel pond in the building containing Reactor Unit 4. This resulted in several disruptive explosions, because overheated zirconium fuel cladding reacted with water and steam and generated a hydrogen cloud which was then ignited. Major releases of radioactivity occurred, initially to air but later via leakage to the sea. The operators struggled to restore full control.

This was a serious nuclear accident, provisionally estimated to be of Level 5 on the International Nuclear Event Scale (INES), a figure which was later amended to a provisional Level 7 (the highest category). The Japanese authorities imposed a 20 km radius evacuation zone, a 30 km sheltering zone, and other countermeasures. Governments across the world watched with concern and considered how best to protect those of their citizens who were residents in Japan from any major radioactive release that might occur (Weightman, 2011).

Some have commented on reports of plant damage caused by the earthquake, concluding that the loss of effective cooling for the reactors stemmed directly from the earthquake rather than the subsequent tsunami. However,

Learning from Failures. DOI: http://dx.doi.org/10.1016/B978-0-12-416727-8.00009-6

the information available on the emergency cooling systems and analysis of the circumstances does not support such a hypothesis (Weightman, 2011).

This case study is a good example of a double-jeopardy, where the combination of earthquake and tsunami caused destruction on a scale that was not anticipated in the initial design specifications. For example, the plant was protected by a seawall—designed to withstand a tsunami of 5.7 m (19 ft), but the wave that struck the plant on March 11 was estimated to have been more than twice that height, at 14 m (46 ft). This, coupled with the now reported land movement of 2.4 m experienced by much of Japan, ensured that the tsunami caused enormous damage along the coast (IMechE, 2011).

9.1.2 The Consequences of the Failure

The earthquake occurred under the sea near the northeast coast of Japan. It lasted over 90 s and caused widespread damage to property, although, due to the civil building design standards most properties did not collapse. As a result of the earthquake, Japan has moved 2.4 m laterally and dropped 1 m vertically. Also, the earth's axis has moved 0.17 m and the length of the earth's day is now shorter by 1.8 μm (IMechE, 2011). This was by any measure a major global event.

The earthquake produced a tsunami 14 m high that struck the coast of Japan and traveled up to 10 km inland, devastating infrastructure already weakened by the earthquake. There were approximately 15,000 confirmed deaths and 10,000 people remain missing. It has been reported that the accident eventually cost Japan between 5% and of its GDP, or US$300−600 billion (Kashyap et al., 2011).

The infrastructure affected included many different types of facility, such as houses, hospitals, electricity and water supplies, petrochemical and oil installations. However, it can be argued that the most significant damage in a global context was to the Fukushima Nuclear Power Station at the town of Okuma. Fukushima is a city in the Tohoku region of Japan. It lies 250 km north of Tokyo, covering an area of 746.43 km^2. As of May 2011, it had a population of 290,064.

The damaged caused by the earthquake and subsequent tsunami, which arrived at 15.41 JST (Weightman, 2011), resulted in mandatory evacuation of the population within a 20 km radius around the site, loss of containment of radiological material to air, contamination in the sea (since detected in the Irish Sea) and of drinking water in Japan.

9.1.3 **The Japanese Nuclear Industry**

Japan is heavily dependent on its nuclear industry, with 54 nuclear reactors currently in operation consisting of 30 boiling water (BWR) and 24 pressurized water (PWR) reactors. The industry is regulated by the Nuclear Safety Commission (NSC) through the Nuclear and Industrial Safety Agency (NISA), which are accountable to the government through the Ministry of Economy, Trade, and Industry (METI) (Weightman, 2011). It was the stated goal of the Japanese government, prior to this event, is that 50% of their electrical power should be nuclear power (although this, of course, may not continue to be the case). In the short to medium term, the Japanese government has suspended operations at Tohoku until the sea defenses are improved, which is estimated could take years to complete.

In an article in the *Guardian Newspaper* (The Guardian, 2013), Mr. Naomi Hirose, president of the Tokyo Electric Power Company (TEPCO), which runs the stricken Fukushima plant, said *"nuclear managers should be prepared for the worst in order to avoid repeating Japan's traumatic experience,"* and then he continues to say *". . . we have to keep thinking: what if."* Hirose said that *"although the situation facing Fukushima Daiichi on 11 March was exceptional, measures could have been adopted in advance that might have mitigated the impact of the disaster. Tepco was at fault for failing to take these steps."* According to him, *"preventative measures included fitting waterproof seals on all the doors in the reactor building, or placing an electricity-generating turbine on the facility's roof, where the water might not have reached it. In addition, wrong assumptions were made,"* he said. Finally he concluded with the following lesson: *"What happened at Fukushima was, yes, a warning to the world,"* he said. The resulting lesson was clear: *"Try to examine all the possibilities, no matter how small they are, and don't think any single countermeasure is foolproof. Think about all different kinds of small countermeasures, not just one big solution. There's not one single answer. We made a lot of excuses to ourselves . . . Looking back, seals on the doors, 6 one little thing, could have saved everything."*

9.1.4 **Some Basic Information About Risk Assessment in Nuclear Industry**

The International Nuclear and Radiological Event Scale (INES) was introduced in 1990 (*World Nuclear News*, 2013) by the International Atomic Energy Agency (IAEA) in order to enable prompt communication of safety significant information in the event of nuclear accidents.

The selection of a level, on the INES, for a given event is based on three parameters: whether people or the environment has been affected; whether any of the barriers to the release of radiation have been lost; whether any of the layers of safety systems are lost. Broadly speaking, events with consequences only within the affected facility itself are usually categorized as "deviations" or "incidents" and set below scale or at levels 1, 2, or 3. Events with consequences outside the plant boundary are classified at levels 4, 5, 6, and 7 and are termed "accidents."

The scale is intended to be logarithmic, similar to the movement magnitude scale that is used to describe the comparative magnitude of earthquakes. Each increasing level represents an accident approximately 10 times more severe than one on the previous level. Compared to earthquakes, where the event intensity can be quantitatively evaluated, the level of severity of a man-made disaster such as a nuclear accident is more subject to interpretation. Because of this, the INES level is assigned well after the incident of interest occurs. Therefore, the scale has a very limited ability to assist in disaster-aid deployment.

Nuclear reactor incidents/accidents are classified using the following scale (in descending order of criticality):

> 7—Major accident (Chernobyl, 1986—USSR, and Fukushima, 2011—Japan)
> 6—Serious accident
> 5—Accident with wider consequences (Three Mile Island, 1979—USA)
> 4—Accident with local consequences (Windscale, 1957—UK)
> 3—Serious incident
> 2—Incident
> 1—Anomaly
> 0—Below scale/no safety significance.

Note that up to level 3 on this scale, the event is classified as an incident, whereas from level 4 onward the event is classified as an accident (Figure 9.1).

Radiation terminology in terms of amount absorbed is measured in RAD (radiation absorbed dose) or in the newer system internationale (SI), using the term Gy (gray). A more important measurement is radiation damage measured in REM (radiation equivalent mammal) or the new term Sv (sievert). Some guidelines:

> 1 sievert = 100 REM (a very serious dose)
> 0.01 (centi) sievert = 1 REM (not a serious dose, but should be avoided)

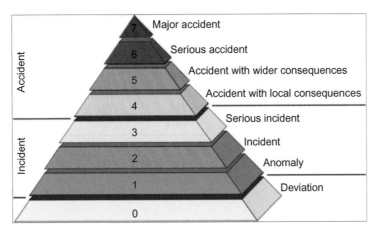

■ **Figure 9.1** The INES scale of nuclear accidents.

0.001 (milli) sievert = 0.1 REM (not a serious dose)
0.000001 (micro) sievert = 0.0001 REM (not a threat)
400 millisieverts = 40 REM (a dangerous but not lethal dose)

Radiation damage potential is a function of intensity and time.

9.2 **ANALYSIS OF FIRST GROUP OF DELEGATES**

This report discusses the accident and proposes that this the loss of control event is due to poor design, leading to a common mode failure for the critical system (the cooling water) and the installed backup systems, and that there was insufficient provision for alternative cooling water supply or for controlled safe pressure relief. This argument is encapsulated in a fault tree.

9.2.1 **Technical Background**

The nuclear plant complex at Fukushima Daiichi is operated by TEPCO. It consists of 6 BWRs with a combined power output of 4.7 GW (Table 9.1) (Weightman, 2011).

BWR use mixed oxide fuel (MOX) which contains a combination of uranium and plutonium. The fuel is used to produce heat by nuclear fission in the reactor core, where the heat produced is transferred into water to produce steam, the same water also being used to moderate the reaction (see Chapter 7). The steam is conditioned (heated) to ensure it is suitably dry and then used to produce electricity via a turbine/generator. This is a circuit (Figure 9.2), the used steam being cooled, and the resultant water

Table 9.1 Overview of the Fukushima BWR Design Types and Outputs

Unit	1	2	3	4	5	6
Reactor model	BWR-3[a]	BWR-4	BWR-4	BWR-4	BWR-4	BWR-5
Containment model	Mark 1	Mark 1	Mark 1	Mark 1	Mark 1	Mark 2
Electrical output (MWe)	460	784	784	784	784	1100
Commercial operation	1971	1974	1976	1978	1978	1979

[a]Fukushima-1 Unit 1 is an early BWR-3 model that has a number of features of the earlier BWR-2 model.

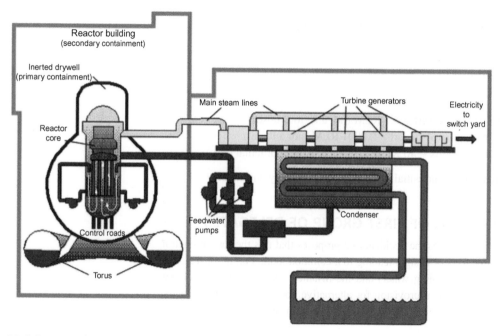

■ **Figure 9.2** Cooling system schematic for a BWR reactor. *Weightman (2011).*

returned to the reactor to be reheated via variable speed electrically cooling water pumps (Areva, 2011).

When the earthquake struck, Reactors 1, 2, and 3 were operational and at full output. Reactors 4, 5, and 6 were not online but were shut down for routine periodic inspection.

9.2.2 **The Cooling Water System**

Irradiated nuclear fuel is self-heating; as the products of nuclear fission decay they release energy in the form of heat. It follows that the safe operation of nuclear BWR is dependent on the provision at all times of adequate cooling water.

Even when a reactor is shut down and the nuclear fission chain reaction stopped, the products of the nuclear fissions will be decaying radioactively and producing heat. So, if the cooling water flow is interrupted, the water will gradually heat to the point of boiling and evaporate. If this is not corrected the fuel rods, in the reactor and in the spent fuel storage pond, will eventually be uncovered, will overheat and explosive hydrogen will be produced by the reaction of steam and water with the hot zirconium fuel cladding. Finally, the fuel rods will melt, distort, and the melted material slump to the bottom of the reactor.

Nuclear facilities which are dependent on cooling water for safe operation have defense in depth to ensure operational resilience. The first line of defense is the water driven by the variable speed pumps powered by the national grid (i.e., off-site) electricity supply. These pumps have operational redundancy in the form of spare pumping capacity.

If grid power becomes unavailable, sufficient AC alternative generation to power all critical equipment is provided by diesel generators connected to the electrical distribution system (which constitutes a second line of defense). In addition, battery backup DC power supply provides a third line of defense, although this DC power continues only for as long as the batteries last (8 h in this case).

9.2.3 **Overview of the Disaster**

Only Reactors 1, 2, and 3 were operational and at full output at the time of the earthquake (IMechE, 2011). They successfully shut down automatically when the earthquake was detected but, as explained, the reactor cores continued to produce heat, from the decay of fission products, which needed to be removed by continuing cooling water flow.

Due to the earthquake, electricity supply to the plant was lost, which led to the emergency diesel generators starting. However, when the tsunami hit Fukushima, the generators, which were situated below ground level in the lower turbine hall (Weightman, 2011), stopped suddenly as they were overwhelmed by the water surge estimated to be between 4 and 6 m in height. The plant was now without AC power and reliant on the battery backup provision.

Even though replacement diesel generators were obtained and available after 13 h (IMechE, 2011), the connection point to the electrical distribution system was flooded and they were not usable. Eventually, the battery power was depleted and the plant was without any electrical power at all.

The consequence of no power was that the ability to provide cooling water to the plant, required for safe shutdown, and using the installed pumps, was lost. The cooling water present in both the reactor cores and the spent fuel pond was therefore stationary and heating led to its evaporation, insufficient level in the reactors, and in the spent fuel ponds, and increased pressure in the system. In the presence of radiation water suffers from radiolytic decay which evolves hydrogen, which under normal circumstances (i.e., flow) is removed from the circuit. Under zero flow conditions, however, it builds up. More significantly the fuel was now exposed (not under water). This led to overheating of the zirconium cladding, hydrogen evolution (as explained), and migration of radioactive species no longer contained by the cladding into the cooling water system (Weightman, 2011).

In order to protect the reactor integrity, a decision was made to release the pressure in the cooling water system. However, the hydrogen now entrained in the cooling circuit was at a level such that when the pressure was released, sufficient hydrogen was present to fuel an explosion which caused significant damage to the reactor buildings and released radioactivity to the atmosphere.

Attempts to provide alternative cooling continued, including the use of:

- Firefighting equipment/water cannons to spray the building exterior (Weightman, 2011).
- Helicopters to drop water.
- Portable pumping systems to supply seawater, via a fire hose connection, to the cooling circuit. However, the supply rate was limited to ~ 2 m^3/h due to limited connection size (Weightman, 2011).

It is known that, for example, the volume of water sprayed into the spent fuel ponds was greater than the volume of the pond (Weightman, 2011), and hence it is clear that during the disaster the radiological containment integrity was degraded. Subsequently, radiological measurement has confirmed that contamination spread to the atmosphere and then to the sea, and subsequently into drinking water supplies.

9.2.4 **Analysis of Contributory Factors**

The full picture of the events at Fukushima has yet to become apparent. However, the following is a discussion of the contributory factors which also impacted (sometimes positively, sometimes not) on cooling, and hence on nuclear safety and plant integrity.

9.2.4.1 *The Basis of the Design*

The design of Reactor Unit 1, which was the oldest design and suffered the most damage, was based upon a 0.18 g (1.74 m/s^2) earthquake, which is equivalent to one of a magnitude 7.5. The March 11 earthquake was much more severe at 0.52 g (5.07 m/s^2), or magnitude 9.0, which in terms of energy is a factor of 1015 greater than the design basis event. But, due to the factors of safety inbuilt in the design, the civil structure survived (IMechE, 2011).

Regarding a tsunami, the design basis was for a 6 m wave. The actual wave was 14 m. An internal to TEPCO safety review, in 2007 (around 4 years prior to the accident), identified that the occurrence of a wave that could overtop the seawall was a 1 in 50 year event (Krolicki et al., 2011). Fukushima 1 was under construction in 1967 (44 years ago), and therefore the 1 in 50 year estimate was both accurate and equivalent to a once in a plant lifetime event. The sea defense, however, had not been improved, and actually fell by approximately 1 m during the earthquake.

9.2.4.2 *The Cooling Water System*

The mains power supply was lost during the initial earthquake. Typically, such supply is duplicated, protected, and provided via diverse routes to allow for catastrophic damage. It is unclear whether this was the case at Fukushima, i.e., were there two routes and, if so, why were they both allowed to be susceptible to the same common mode failure?

Initially, the diesel generator system was successfully used when the earthquake led to a loss of mains power. However, the location of both the generators and the associated distribution board (below ground level) and the lack of appropriate protection allowed the tsunami to overwhelm and damage them beyond use. This could have been avoided by simply locating them above the height of any possible tsunami surge.

There was no duplicate or alternative connection point for any additional diesel generators, as evidence by the fact that some 13 h later generators were available but not connectable. In fact it was deemed that repairing the mains supply would have been the quickest method of reinstating the pump power supply (IMechE, 2011).

No evidence has been identified regarding whether the cooling water systems from the other reactors were cross-connectable, i.e., could the cooling water pumps, generators, etc. from adjacent reactors (in cold shutdown and therefore with spare capacity) be used to cool an uncooled reactor?

There were difficulties in providing seawater to the cooling system while the cooling water pumps were unavailable. There were deficiencies in the preplanning and readiness of the workforce to use the temporary pumping system commissioned to pump via hose connection points. Due to the limits of physical size of the connection, the maximum throughput was insufficient to achieve the required cooling rates (Weightman, 2011).

Due to insufficient cooling the pressure rose to unacceptably high levels requiring venting to atmosphere. That this was required indicates that the system either had no or not enough installed pressure relief valves, and that any blowdown was not contained in radiological containment.

9.2.4.3 *The Emergency Response*

There were deficiencies in the preplanning and readiness of the workforce to use the temporary pumping system commissioned to pump via hose connection points.

9.2.5 **Fault Tree**

It is proposed that the process of evolution of the hydrogen explosion above Fukushima can be represented by a fault tree, as shown in Figure 9.3.

9.2.6 **Discussion of First Group of Delegates**

The Fukushima Daiichi incident was fundamentally down to poor design, in that the normal mains and both backup power supplies were allowed to fail due to a single common mode failure, albeit from an extreme natural event. The mains supply integrity was such that the earthquake damaged it beyond repair and no diverse supply remained intact. The diesel generator system was located in a plant room likely to be swamped, and again no diverse connection point remained. And finally, neither could be repaired before the backup power supply was exhausted. The potential for an earthquake-generated tsunami in excess of the existing sea defenses, and therefore capable of these effects, was not only realistic, it was actually foreseen in 2007 and calculated as likely in the lifetime of the plant (Krolicki et al., 2011). But the plant continued to be operated and the sea defenses were not improved, and the resilience of the cooling water system was not increased.

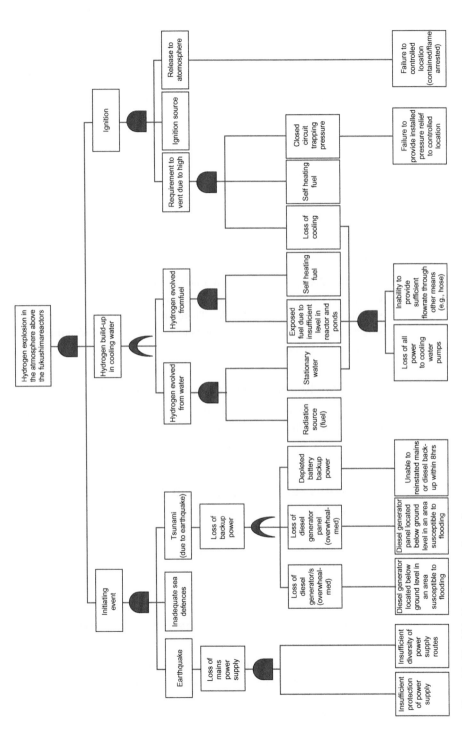

■ **Figure 9.3** FTA of Fukushima disaster according to first group.

Furthermore, the ability to provide cooling by other means was insufficient both with respect to the training and readiness to do so, and also as regards to the physical hardware to do so. Quite simply, without the installed pumps, they could not provide enough water to prevent the pressure rising to unacceptable levels by any other installed emergency system (Weightman, 2011).

The ability of the installed design to control, contain, and direct excessive pressure was insufficient. Even if cooling could not be reestablished, there should be a means of safely directing the vented material to a suitable location. This was absent. In addition, the vent was not controlled from the point of view of fire suppression. In both cases, the venting to a suitably large installed system containing, for example, nitrogen blanketing would have limited the potential for the vented material exploding.

Given that the possibility of such a large tsunami was foreseen (Krolicki et al., 2011), it follows that through that the consequences were also foreseeable via a suitable FMEA. Therefore, the failure to carry out suitable hazard analysis, and implement the actions thus identified, was also a design failure.

9.2.7 Conclusion of First Group of Delegates

The event was foreseen and design shortcomings were not investigated nor addressed. This aspect of the disaster, the hydrogen explosions, was fundamentally due to the lack of resilience of the cooling water circuit.

9.3 ANALYSIS OF SECOND GROUP OF DELEGATES

9.3.1 The Analytic Hierarchy Process

In assessing the possible course of action following the incident, the analytic hierarchy process (AHP) technique will be used to weigh the multiple options proposed regarding Japan's power dilemma created by the recent debate on nuclear power safety.

9.3.2 Design Evaluation of the Fukushima Nuclear Plant

Design of nuclear power plants is done with due consideration of the associated risks with, the consequences being intrinsically on the high side, implying that it is the likelihood of failure that has to be minimized. The Fukushima plant was no exception in this respect.

9.3.3 **BWR Design for Mitigation of Abnormal Conditions**

In the event of an abnormal condition, the engineered safety system had to be able to fulfill three fundamental functions, namely:

a. Safe shutdown of the reactor. When any anomaly occurred during reactor operation, the reactor had to shut down to prevent further nuclear fission. This is accomplished by inserting neutron-absorbing control rods into the reactor core.

b. Removal of the heat from radioactive decay. As explained, after the nuclear fission is stopped, the reactor releases heat from the radioactive decay of fission products. Continued cooling of the reactor after shutdown is therefore necessary to avoid over-heating and the resulting fuel damage.

c. Confinement of radioactive material within the containments. If radioactive material leaks from the core, it must be prevented from leaking into the environment. Containment is the ultimate means of protecting the public against radiation exposure (JNES, 2005).

9.3.4 **The Cooling System Design**

9.3.4.1 *Power Supply*

As explained, cooling systems for nuclear power plants are multi-redundant. Apart from the redundancy of the electrically driven pumps themselves, the power supply is also designed with a great deal of redundancy. At Fukushima, the four interconnected external power lines were supplemented by diesel-powered engine generators interconnected through a network of station bus bars. A schematic of the power grid across the plant is shown in Figure 9.4.

Station blackout would entail all power systems (internal and external) failing, or the switchgear failing. A fault tree representation of the event *station blackout accident* for the single line interconnection diagram of Figure 9.5 is shown. Note that the grid does not show the DC system which is shown on the fault tree diagram.

Also note that the three undeveloped events P2, P3, and P4 depend on system redundancy as is the case with Unit 1, i.e., two diesel-powered AC generators plus a DC power supply system. An RBD of the complete power system arrangement would show that it was a highly reliable system.

The undeveloped event SW, the switching system, was also a highly redundant arrangement as can be seen from the network interconnections as shown

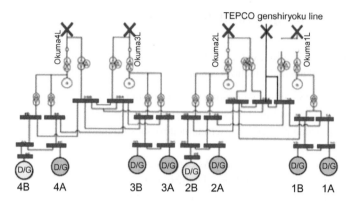

■ **Figure 9.4** Single line diagram of Units 1—4, Fukushima Daiichi, TEPCO Interim Report, December 2011.

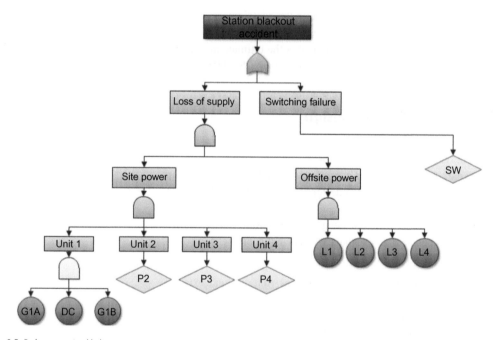

■ **Figure 9.5** Fault tree: station blackout event.

in Figure 9.4. The full diagram for the switching system is not shown, but it will suffice to say that there was great deal of redundancy in the design.

9.3.4.2 *Pump Redundancy and Diversity*

In the event that there was a common mode of failure of the highly redundant power supply system for the electrically driven pumps, the stations were also equipped with steam-turbine-driven pumps, and all these pumps were inter-connected with crossovers in such a way that the likelihood at loss of cooling due to pump failure in any of the four units was greatly minimized.

9.3.4.3 *Redundant Water Sources*

As a protection against failure of storage of the water needed for cooling, through accidental chemical contamination, for example, or dry out, there was also redundancy of the cooling water storage arrangements.

9.3.5 **Design Against Earthquakes and Tsunamis**

Establishing permits for the Fukushima power plants were granted in the years between 1966 and 1972. During that time Japan had no site design guidelines for defense against tsunamis. For Fukushima, a design height of OP* + 3.122 was adopted, based on historical records of the highest earthquake tidal wave level measured, i.e., after the 9.5 magnitude Chile earthquake of 1960 (information provided by TEPCO relates heights of both the tsunami and the seawall to a level known in Japan as OP in a similar manner to which Ordnance Survey maps in the UK are referenced to sea level). OP is the baseline level known as the Onahama Port Base Level. At that time this was considered as the worst case.

In 2002, the Japan Society of Civil Engineers (JSCE) released a report, *The Tsunami Assessment Method for Nuclear Power Plants in Japan*. This recommended revised tsunami preparedness measures based on new find-ings. The report findings were consistent with earthquake level predictions from the *Headquarters for Earthquake Research Promotion*, which predicted possible occurrence of an earthquake of approximately 8.2 magnitude in the vicinity of the region. The Fukushima Nuclear Power Plant (NPP) tsunami design basis was revised to OP + 5.7 in line with the recommendations.

Further studies were being undertaken by TEPCO after this initial review, in liaison with the Headquarters for Earthquake Research Promotion and the JSCE to ensure conformity of the nuclear sites to any new findings from tsunami wave source models.

9.3.6 **Exploring the Incident**

Having looked at the original design and subsequent improvements in the defenses against abnormal conditions, the question that comes to mind is whether the Fukushima disaster could have been avoided, either during the initial design or after the subsequent improvements. A look at the sequence of events immediately prior to the incident is informative in this respect.

March 11, 2011, at 14.46 Japan time, the earthquake (magnitude 9) struck of the coast of Japan. The Great East Japan Earthquake (as it was later called) generated a series of large tsunami waves that struck the east coast of Japan. As well as other enterprises, several nuclear power facilities were affected by the severe ground motions and large multiple tsunami waves. The affected stations were Tokai Dai-ni, Higashi Dori, Onagawa, and TEPCO's Fukushima Daiichi and Dai-ni.

The operational units at these facilities were successfully shut down by the automatic systems installed, as part of the design, to detect and respond to earthquakes. However, the subsequent large tsunami waves degraded all these systems to varying degrees, with the most serious consequences occurring at Fukushima Daiichi.

Although all off-site power was lost when the earthquake occurred, the automatic systems at Fukushima Daiichi successfully inserted all the control rods into its three operational reactors upon detection of the earthquake, and all available emergency diesel generator power systems were put into operation, as designed. The first of a series of large tsunami waves reached the site about 46 min after the earthquake.

These tsunami waves overwhelmed the defenses of the plant, which were only designed to withstand waves of a maximum of 5.7 m high. The larger waves that impacted on this facility on that day were estimated to be over 14 m high. The tsunami waves reached areas deep within the units, causing the loss of all power sources except for one emergency diesel generator (6B), with no other significant power source available on or off the site, and little hope of outside assistance (Seneviratne, 2011).

The station blackout and the impact of the tsunami caused the loss of all instrumentation and control systems at Reactors 1−4, with Emergency Diesel 6B providing emergency power to be shared between Units 5 and 6.

The tsunami and associated large debris caused widespread destruction of many buildings, doors, roads, tanks, and other site infrastructure, including loss of heat sinks. The operators were faced with a catastrophic, unprecedented emergency with no power, reactor control, or instrumentation, and in addition, severely affected communications systems both within and external to the site.

With no means to confirm the parameters of the plant or cool the reactor units, the three reactor units that were operational up to the time of the earthquake quickly heated up due to the usual reactor decay heating. There were efforts, by the plant personnel, to contain this otherwise catastrophic situation, but inevitably there was severe damage to the fuel and a series of explosions occurred, which caused further destruction at the site, making the situation even more hazardous for the plant personnel. Subsequently, a considerable level of radiological contamination spread into the environment. The IAEA rated the incident as of Level 7 severity, the highest rating on the INES (Seneviratne, 2011).

9.3.7 **The Nuclear Safety Debate**

Exploration of the Fukushima incident leads back to the question *"What went wrong?"* Could the station blackout have been avoided? Was it an engineering design and operations problem or a management and regulatory system failure?

A Greenpeace International report (Morris–Suzuki, 2012) on the incident claimed that the accident marked the end of what it called the *"nuclear safety" paradigm*. The report drew the unusual conclusion that the notion of *nuclear safety* does not exist after what happened at Fukushima, but all that can be talked about concerning nuclear reactors are risks, unknown risks in the worst case. The report went on to say that, at any time, an unforeseen combination of technological failures, human errors, or natural disasters at any one of the world's reactors could lead to a reactor quickly getting out of control.

The report questioned the defense in depth of the engineering design barriers for nuclear power plants and disputed the probabilistic risk assessment (PRA)-based postulation of only one core meltdown likely to occur in every 250 years. Being a humanitarian-focused organization, Greenpeace did not consider the technicalities leading to the Fukushima accident, but rather focused on the response both by the licensed operator, in this case TEPCO, and the Japanese regulatory authorities. It did not spare the IAEA in laying the blame and flaws on the agency's stance on the incidence. What becomes clear, one of the contributors to the report claimed, is that the weaknesses in the regulation and management of Japan's nuclear power industry have not been "hidden" faults in the system. On the contrary, people had been aware of, written, and warned about them for decades (Morris–Suzuki et al., 2012).

So, from the humanitarian viewpoint of the report, the Fukushima accident was a regulatory system failure. Risks were known but no action was taken to address them. From a neutral perspective, this does not justify the claim that safety in nuclear stations is nonexistent. Rather, it points to the need to address some system deficiencies and suggests improvements that can make nuclear power even safer.

The IAEA report, on the other hand, conceded that Fukushima was an extremely unprecedented case and claimed that the response was the best that could be achieved considering the circumstances. However, it accepted that there were insufficient *defense-in-depth* provisions for tsunami hazards, in the sense that although these were considered in both the site evaluation and the design of the Fukushima Daiichi NPP, and the expected tsunami height was increased to 5.7 m after 2002, the tsunami hazard was actually underestimated. However, the view is that this was just a *black swan event* and does not invalidate the applicability of PRA postulates in nuclear power applications.

In the Fukushima case, the additional protective measures taken as result of the evaluation conducted after 2002 were not sufficient to cope with the high tsunami run-up values and all associated hazardous phenomena. What comes out clearly from the IAEA report is that the design review underestimated the tsunami effect and this could therefore be classified as a design and reengineering failure.

The nuclear authorities generally differ regarding the humanitarian view, in the sense that they see the incident as offering an opportunity for improvement in nuclear power PRA, rather than a trumpet for propagating the message that nuclear power should be scrapped or be perceived as a public hazard. The general consensus at the *World Nuclear Fuel Cycle 2011 Conference* was in support of this view, the prevailing view at the conference seeming to be that nuclear energy will be providing utility power around the world for a long time, despite the accident at Fukushima Daiichi. This assertion was based on expert knowledge with minimal application of the decision-making tools available at the time.

9.3.8 **The Nuclear Power Decision for Japan**

Faced with the foregoing two opposite views regarding the place for Japan's (and ultimately the world's) nuclear power usage, we shall now explore the available options for human safety-driven improvement (or change) applicable to the Japan circumstances with respect to utility power after the Fukushima incident.

9.3.8.1 *Replace All Nuclear Power with Alternative Sources*

This is a popular view among the environmental protection and humanitarian organizations. The Greenpeace report suggested that a significant nuclear accident is bound to occur every decade, based on known incidences, and that puts a question mark over the applicability of nuclear power from the environmental safety perspective. The option to replace all NPPs in Japan is, however, based on the assumed existence of renewable energy sources, or other safer alternatives that could make up for the nuclear phaseout.

9.3.8.2 *Continue Using NPP with Improved Barriers to External Influences and Better Legislation*

This is a popular view among the nuclear industry professionals. It is based on the belief that nuclear is one of the safest forms of energy and that PRA postulates on the probability of occurrence of catastrophic nuclear accidents are generally correct, i.e., the probability is remote. Failure in this case is an opportunity for learning, albeit that it comes at a great cost. Others have gone as far as proposing a review of how sites for nuclear power plants are selected by considering the historically based probability of natural occurrences.

9.3.8.3 *Continue with the Status Quo*

This option is based on the view that nuclear accidents of large magnitude are *black swan* incidents, 1 in every 250 years according to present PRA theory. This *black swan* claim however would appear to the environmental pressure group to be undermined by the much shorter time lapse between the Chernobyl and Fukushima disasters.

9.3.9 **Application of MCDM**

The three available options here are subjected to an Multiple Criteria Decision Making (MCDM) process, namely, AHP based on the attributes of safety, environment, economy, image, and feasibility. The image criterion is considered from the legislature's point of view, i.e., that of the Japanese government and its nuclear regulatory agency, the Nuclear Industrial Safety Agency (NISA) and, on the extreme end, the IAEA and its affiliates. The AHP hierarchy thus developed is shown in Figure 9.6.

9.3.10 **AHP Results**

Following traditional AHP guidelines, the five attributes in the hierarchy were weighted, between 0 and 1. The attribute "*feasibility*" has the highest score, whatever alternative is to be chosen; first and foremost the alternative has to be feasible, then the rest can be considered, otherwise the analysis would be of no practical use and a waste of resources. Subsequent pairwise comparisons were done, more importantly the one with alternatives for the opposing sides, i.e., environmentalists and IAEA.

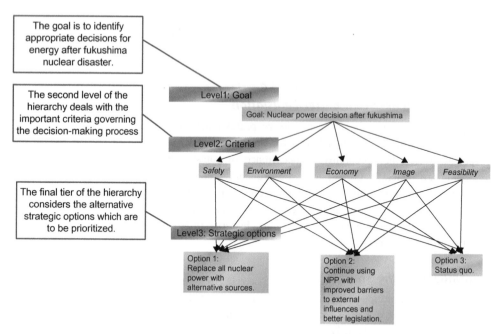

■ **Figure 9.6** AHP hierarchy.

The representation of the hierarchical model is shown in Figure 9.7 and it represents priorities given to criteria and alternatives (options). It shows the option "*enhance nuclear safety*" to be the preferred option.

The gradient sensitivity with respect to the attribute "*safety*," shown in Figure 9.8, indicates that, as far as safety is concerned, the use of alternative energy sources is the preferred option, while disregarding safety would result in the status quo option being preferred. This can be illustrated by moving the vertical line that indicates the importance of "safety" to the left, then the highest intersecting option (most preferred) becomes the one that belongs to the option status quo.

9.3.11 **Conclusion of the Second Group of Delegates**

This group applied the *AHP* to decide on the future of nuclear power usage in Japan following the Fukushima accident. Their study is further proof of the applicability of AHP in multiple criteria decision-making processes.

The favored alternative is to continue using nuclear power in the foreseeable future, but with enhanced safety features, derived from revised Probabilistic Risk Assessment (PRA) / Probabilistic Safety Assessment

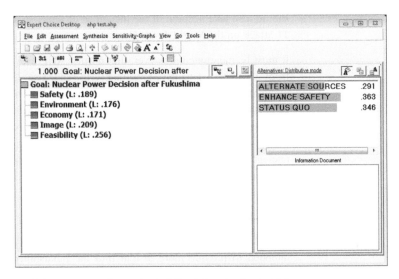

■ **Figure 9.7** Hierarchical model showing priorities of criteria (on the left) and alternatives (on the right).

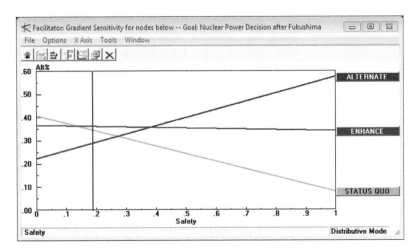

■ **Figure 9.8** AHP sensitivity graph with respect to "safety."

(PSA), to deal with the advent of extraordinary forces of nature such as the one that devastated Japan in March 2010. The concept of continuous improvement touted by proponents of *PRA* should be the guiding principle. One proposal is to possibly reconsider tectonic characteristics for nuclear power sites. Another is that the defense-in-depth structural design should also take account of the incidence of terrorist action such as the 7/11 attack on the World Trade Center.

It must be mentioned, however, that normal AHP uses aggregate analyses from a number of people, presumably to reduce subjectivity, include all relevant stakeholders and promote consistency. But this has not been the case with this study as only one small group of participants was employed to carry out the analysis. Nevertheless, it does provide a framework which could be used by a number of participants to settle the nuclear power debate. The use of *expert knowledge* as prescribed by AHP could add more credibility to the findings of the analysis.

9.4 FEEDBACK AND GENERIC LESSONS

Here the tutor (the present author) offered feedback to the two groups, intended to provide more insight into the Fukushima accident, into the company and culture, and hence addressing various specific and generic issues.

9.4.1 Overall Criticism of the Two Groups

The groups produced slightly different mental models, even though they were exposed to the same narrative. The main differences were in the level of detail each group went into with respect to the fault tree analysis (FTA), the first group offering more insight into the technical issues, the second group using strategic qualitative analysis in the AHP approach. Nevertheless, on the whole, there was more agreement than otherwise. A criticism of the first group's FTA is that there was a confusion between cause and effect; events such as tsunami and earthquake should have been modeled as basic events and hence should have been positioned at lower, rather than upper, levels of the fault tree. A proposed enhanced FTA is shown in Figure 9.9.

As shown in Figure 9.9, the hydrogen explosion and the meltdown were due to three simultaneous factors: loss of coolant (ultimate heat sink), hydrogen built-up in cooling water, and ignition (or detonation), where the term "ultimate heat sink" refers to the function of dissipation of residual heat after a shutdown or an accident.

9.4.2 Wider Generic Lessons for the Nuclear Power Industry

Across the world, there are more than 400 nuclear power reactors operating, with over 140 in Europe, 54 in Japan (Weightman, 2011), and around

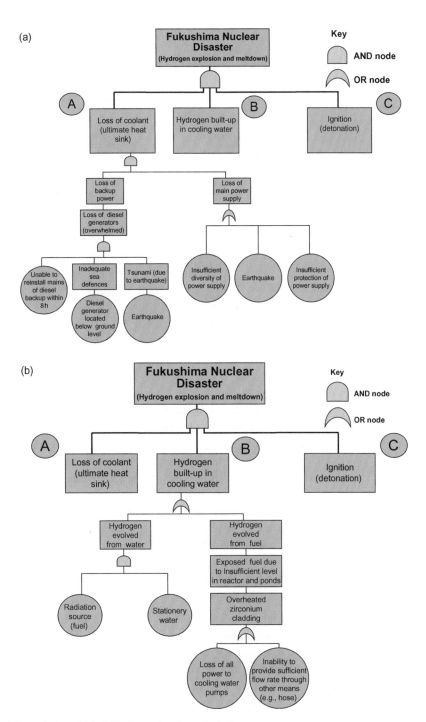

■ **Figure 9.9** (a—c) Proposed enhanced FTA of Fukushima nuclear disaster for the first group.

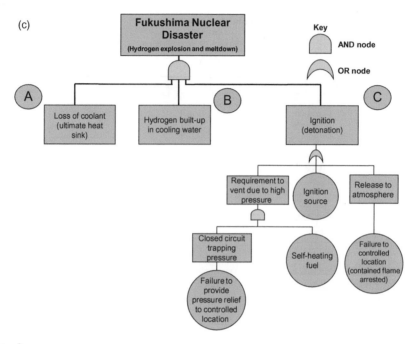

(c)

■ **Figure 9.9** (Continued)

100 in the United States. Fukushima is the third major nuclear accident (i.e., it was preceded by Three Mile Island and Chernobyl), and all three happened within less than half a century, which makes us question our models and original assumptions. To test for a cumulative probability of a 1 in 10 million chance of a nuclear failure each year would require building 1000 reactors and operating them for 10,000 years. Clearly, Three Mile Island, Chernobyl, and Fukushima each arose from very different circumstances, invalidating various modeling and risk assessment assumptions, and resisting assimilation into a single data set. It is difficult, with such a small sample size, to make generalizations about where current risk models fail, though it would suggest that the original ambitious annual failure risk estimates were serious underestimations.

As with the accidents at Three Mile Island and Chernobyl, the lessons tend to fall within one of two categories: those that blame the technology (such as the reactor design) and those that blame social factors (such as poorly conceived regulation or corporate greed) (Pfotenhauer et al., 2013). However, nuclear safety is not simply a matter of developing more powerful risk models and more accurate worst-case scenarios. It is equally about developing better strategies for enabling societies to understand the role of modeling and deal with its inescapable limitations. The modeling of complex sociotechnical systems itself incorporates sociotechnical assumptions that modelers and decision makers need to keep in mind (Pfotenhauer et al., 2013).

The time has come to separate national economic and political interests in promoting nuclear power from the regulatory function, which concerns all nations. Recently recommended has been the elevation of the mandate of the IAEA to include a licensing function for nuclear power plants, thereby changing its status from an advisory body to that of an international institution with authority to make legally binding decisions (Pfotenhauer et al., 2013). Such a move would go beyond other suggested expansions of the IAEA mission in light of Fukushima, such as the inclusion of disaster response mechanisms. Under this new international regulatory regime, a global community would be involved in decisions about the siting of nuclear power plants, the proliferation of reactor technology, safety enforcement, and financial responsibility for events such as Fukushima. Nuclear disasters do not respect national borders, and governance models must therefore include a significant global dimension.

In an article titled "Learning lessons from Fukushima," Mike Weightman, who was the UK Chief Inspector of Nuclear Installations, and head of a fact finding mission by the IAEA, concluded that *"The Magnitude-9 earthquake caused severe ground motions that lasted for several minutes at the Daiichi plant. The measured motions reasonably matched the predictions of the designers of the seismic protection measures. Upon detection of these ground motions, the safety systems at Daiichi shut down the reactors and started the backup systems. All the evidence I have seen, including from the other Japanese nuclear power plants that witnessed similar ground motions, supports the view that the Daiichi plant safely survived this massive earthquake. However, the flood protection measures at the Daiichi plant were originally designed to withstand a 3.1 m high tsunami, whereas the largest wave that crashed into the site in March inundated it to around 15 m. A review in 2002 by the operators of the Daiichi plant did result in increases to the tsunami defenses to enable it to better survive a 5.7 m high tsunami. This improvement still proved to be inadequate, especially considering the history of tsunamis along that coast over the past century"* (Weightman, 2012).

In any process industry, and especially in nuclear energy, one can attribute failure to one of two main reasons: either a design integrity failure or an operational and maintenance failure. Fukushima was clearly caused by the former rather than the latter. Its impact was limited, because of the

Japanese safety procedures to which we should pay tribute. However, it is clear that the design integrity of the plant did not include a sufficient factor of safety.

There is evidence that there were various prior concerns about the design of the Fukushima plant. According to Krolicki et al. (2011):

1. *Evidence of previous research*: A research paper, presented at a nuclear engineering conference in July, 2007 concluded that there was a roughly 10% chance, based on the most conservative assumptions, that a tsunami could test or overrun the defenses of the Fukushima Daiichi nuclear power plant within a 50-year span. But Tokyo Electric did nothing to change its safety planning as a result of that study, which was the product of several years of work at Japan's top utility, prompted by the 2004 earthquake off the coast of Sumatra that had shaken the industry's accepted wisdom. In that disaster, the tsunami that hit Indonesia and a dozen other countries around the Indian Ocean also flooded a nuclear power plant in southern India. That raised concerns in Tokyo about the risk to Japan's 55 nuclear plants, many located on the dangerous coast in order to have ready access to water for cooling.

    Tokyo Electric's Fukushima Daiichi plant, some 240 km (150 miles) northeast of Tokyo, was a particular concern. The 40-year-old nuclear complex was built near a quake zone in the Pacific that had produced earthquakes of magnitude 8 or higher four times in the previous 400 years—in 1896, 1793, 1677, and 1611. Based on that history, Sakai, a senior safety manager at Tokyo Electric, and his research team applied new science to a simple question: What was the chance that an earthquake-generated wave would hit Fukushima? More pressing, what were the odds that it would be larger than the roughly 6 m (18.6 ft) wall of water the plant had been designed to handle?

    The tsunami that crashed through the Fukushima plant on March 11 was 14 m (43.4 ft) high. Sakai's team determined that the Fukushima plant was certain to be hit by a tsunami of 1 or 2 m (3−6 ft) in a 50-year period. They put the risk of a wave of 6 m (18.6 ft) or more at around 10% over the same time span. In other words, Tokyo Electric scientists realized as early as 2007 that it was quite possible a giant wave would overwhelm the sea walls and other defenses at Fukushima surpassing engineering assumptions behind the plant's design that date back to the 1960s.

2. *Responsibility*: Among examples of the failed opportunities to prepare for disaster is that Japanese nuclear regulators never demanded that Tokyo Electric reassess its fundamental assumptions about earthquake and tsunami risk for a nuclear plant built more than four decades previously. In the 1990s, officials urged, but did not require, that Tokyo Electric and other utilities shore up their system of plant monitoring in the event of a crisis.

3. *Assumptions and regulatory rules*: Despite the projection by its own safety engineers that the older assumptions might be mistaken, Tokyo Electric was not breaking any Japanese nuclear safety regulation by its failure to learn from its new research and fortify Fukushima Daiichi, which was built on the rural Pacific coast to give it ready access to seawater and distance it from population centers. *"There are no legal requirements to reevaluate site related (safety) features periodically,"* the Japanese government said in a response to questions from the United Nations nuclear watchdog, the IAEA, in 2008.

   In fact, in safety guidelines issued over the previous 20 years, Japanese nuclear safety regulators had all but written off the risk of a severe accident that would test the vaunted safety standards of one of their 55 nuclear reactors, which were a key pillar of the nation's energy and export policies. That left planning for a strategy to head off runaway meltdown in the worst-case scenarios to Tokyo Electric in the belief that the utility was best placed to handle any such crisis.

4. *Operational rather than strategic concerns*: Japanese regulators and Tokyo Electric instead put more emphasis on regular maintenance and programs designed to anticipate flaws in the components of their ageing plants. Safety regulators, who also belong to METI, did not require Tokyo Electric to rethink the fundamental safety assumptions behind the plant. According to a 2009 presentation by the utility's senior maintenance engineer, the utility only had to ensure that the reactor's component parts were not been worn down dangerously.

   These accounts show the importance of the previous stated recommendations that the mandate of the IAEA should be elevated to include a licensing function for nuclear power plants, thereby changing its status from an advisory body to that of an international institution with authority to make legally binding decisions.

Table 9.2 (*BBC News*, 2011) compares the two nuclear accidents at the Fukushima Daiichi (2011) and Chernobyl (1986) nuclear power plants.

Table 9.2 Comparison of Fukushima and Chernobyl Disasters

Plant name	Fukushima Daiichi	Chernobyl
Location	Japan	Soviet Union (Ukrainian Soviet Socialist Republic)
Date of the accident	March 11, 2011	April 26, 1986
INES level	7	7
plant commissioning date	1971	1977
Years of operation before the accident	40 years	9 years
Electrical output	4.7 GW	4 GW
Type of reactor	Boiling water with containment vessel	Graphite moderated without containment
Number of reactors	6; 4 (and spent fuel pools) involved in accident	4; 1 involved in accident
Amount of nuclear fuel in reactors	1600 tons	180 tons
Cause of the accident	Loss of cooling system due to earthquake and tsunami destroying power lines and backup generators, leading to meltdown. Failure to plan for total loss of off-site power and backup power	Faulty design leading to instability at low power (positive void coefficient), along with poor safety culture, leading to prompt criticality, and steam explosion during an improvized experiment
Maximum level of radiation detected	72,900 mSv/h (inside reactor 2)	200,000 mSv
Radiation released	900 PBq into the atmosphere in March last year (2011) alone up from previous estimates of 370 PBq total. Radiation continues to be released into the Pacific via groundwater, as of September 15, 2013	5200 PBq
Area affected	Radiation levels exceeding annual limits seen over 60 km (37 mi) to northwest and 40 km (25 mi) to south-southwest, according to officials	An area up to 500 km (310 mi) away contaminated, according to the United Nations
Exclusion zone Area	20 km (30 km voluntary)	30 km
Population relocated	300,000	About 115,000 from areas surrounding the reactor in 1986; about 220,000 people from *Belarus*, the *Russian Federation*, and *Ukraine* after 1986 (335,000 people total)
Direct fatalities from the accident	2 crew members (gone to inspect the buildings immediately after the earthquake and before the tsunami) due to drowning	31 (64 confirmed deaths from radiation as of 2008, according to the UN)
Current status	*Cold shutdown* declared on December 16, 2011, but decommissioning will take 10 years	All reactors were shut down by 2000. The damaged reactor is covered by a hastily built steel and concrete structure called the *sarcophagus*. A *New Safe Confinement* structure is under construction and expected to be completed in 2015, from which the plant will be cleaned up and decommissioned

9.5 **CRITICAL COMMENTARY SECTION**

It is noted that not everybody agrees with the findings and conclusions of each chapter. So at the end of some of the chapters there is an alternative view, which is not intended to undermine the issues discussed but rather present other perspectives. The ideas included do not necessarily represent the author's own view, but they are worth consideration and worth initiating some reflection and debate.

CRITICAL COMMENTARY: RISK OF NUCLEAR POWER SUPPLY: IS IT REALLY 1 IN A MILLION PER REACTOR-YEAR?

To definitively test the validity of the assessment that the probability, per year, of a given nuclear power reactor failing (i.e. experiencing a core meltdown) is one in a million would require, at least, the building of 1,000 reactors and operating them for 1,000 years, which would accumulate 1,000 years $\times$ 1,000 reactors = 1 million reactor-years of operation, and then (if the assessment is correct) experiencing only one or two, if any, failures. Across the world there are now more than 400 nuclear power reactors operating (Weightman, 2011). Assuming, as a very approximate working figure that, on average, each of these has been running for, say, 50 years (a considerable over-estimate!) we have now accumulated, at most, a mere 400 $\times$ 50 = 20,000 reactor-years of operation, so if the above risk assessment is anything like correct, it suggests that it should be highly unlikely that any meltdowns at all should have occurred, anywhere.

Environmental groups, however, believe that, in the light of the time lapse between the Three Mile Island, Chernobyl and Fukushima disasters - these three events happening within less than half a century - the nuclear power risk analysts' models and assumptions are highly questionable, to say the least. The three identified accidents arose, however, from very different circumstances, which invalidated various modelling and risk assessment assumptions. It can also be claimed that it is difficult with such a small sample size to make generalizations about where our models fail. The evidence does nevertheless suggest that the original ambitious annual failure risk estimates, such as one in ten million per reactor, or even one in ten thousand, are serious underestimations.

Chapter

10

Hurricane Katrina Disaster

10.1 INTRODUCTION

Previous research has shown that organizations learn more effectively from failures than from successes (Madsen and Desai, 2010) and that failures contain valuable information, but organizations vary in their ability to learn from them (Desai, 2010). It has also been argued that there is a need for a paradigm shift in accidents models due to new challenges that relate to issues such as the fast pace of technological change, the changing nature of accidents, decreasing tolerance to single accidents, and increasing complexity and coupling (Leveson, 2004).

This study of the Hurricane Katrina disaster shows how the reliability techniques of fault tree analysis (FTA), reliability block diagram (RBD), and failure mode effect and criticality analysis (FMECA) can be integrated with the multicriteria decision-making (MCDM) technique in a revealing model of the event.

The contributions of the study are both theoretical and methodological. On the theoretical side, it is shown how data—some of which is based on interpretation and judgment and some is more empirical in nature—can be combined in a rich framework that can be used by decision makers to prioritize different strategies and allocation of resources. On the methodological side, it is shown how the two fields of risk analysis and decision science can be combined and utilized in an integrated manner.

Tinsley et al. (2012), who investigated near-miss events as well as Hurricane Katrina and other disasters, concluded that *"people may be complacent because prior experience with a hazard can subconsciously bias their mental representation of the hazard in a way that often (but not always) promotes unrealistic reassurance."* Here, this work on Hurricane Katrina disaster is extended, by providing tools that can help in performing a systematic analysis that can lead to learning from failures. It is hoped that this sort of analysis will contribute to the provision of a useful mental representation of disasters.

Learning from Failures. DOI: http://dx.doi.org/10.1016/B978-0-12-416727-8.00010-2

Regarding reliability analysis, FTA is employed to show how some of the direct causes and contributing factors of the disaster interacted with each other. RBD is then used to demonstrate how overall system reliability could be calculated and improved through, for example, strengthening weak (series) structures revealed by the analysis. Risk priority number (RPN) is then employed to rank the risk of different failure modes.

Leveson (2004) argued that event chain models have limitations, in their inability to deal with nonlinear relationships where feedback is emphasized. In its MCDM analysis, the work presented here utilizes analytic hierarchy process (AHP) to provide prioritization, sensitivity analysis, and feedback on consistency of the different criteria and the alternative contributing factors. The model helps the decision maker to prioritize different strategies and the allocation of resources. It also provides a sensitivity analysis and a measure of the consistency of feedback. Also discussed are the high-level design improvements, and the lessons learned, which should be acted upon so as to avoid a repeat disaster.

Figure 10.1 outlines the structure and relationship between the different techniques used, where the three techniques FMEA, FTA, and RBD belong to the reliability analysis domain, whereas the AHP is a MCDM technique.

In many ways, disasters are similar to "black swans" a term coined by Taleb (2010) for a phenomenon that has three attributes. First, it is an *outlier*, as it lies outside the realm of regular expectations, an unexpected event. Second, it has an extreme *impact*; it is considered a paradigm shift as it changes the way one looks at things. Third, its occurrence is explainable and predictable, but only after the fact and not before. In summary, the triplets of a black swan, as well as in any disaster, are rarity, extreme impact, and retrospective predictability. Taleb (2010) argued that this phenomenon is accelerating as the world is getting more complicated. The combination of low predictability and large impact makes disasters a great challenge to analyze and hence the work presented here is both timely and valuable. We accept that hurricanes are not that rare, but the Katrina disaster was not just an ordinary hurricane in terms of its impact compared to all previous hurricanes, which made it a unique event.

There are two categories of lessons that can be learnt from the Katrina event which are similar to those proposed by Flouron (2011) when describing the BP Deepwater Horizon accident. The first relates to narrow, or specific, lessons while the second relates to broader issues. The former category arises when describing an event such as a "failure due to

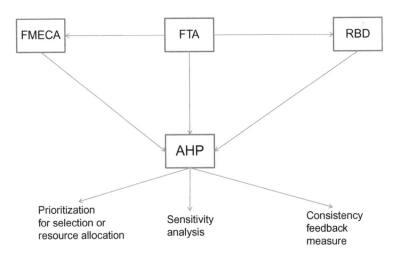

■ **Figure 10.1** The model structure and the relationships between the different techniques.

human negligence, or having insufficiently high, strong, and maintained levees in a hurricane prone area," a sort of technical failure, whereas the latter arises from describing an event such as a "disaster," defined as an occurrence inflicting widespread destruction and distress. Through the latter lens the event is considered broadly in a social science, political, systems theory, and management approach, as suggested by the seminal work of Turner (1978) which was then followed up by Toft and Rynolds (1997).

This kind of analysis can produce four main benefits: First, it can help to identify the root cause of what went wrong and why. Second, it can act as an early warning signal just prior to the event in order to take preemptive measures. Third, it can help to institute long-term plans to prevent similar events from reoccurring. Fourth, it can provide decision makers with a set of priorities for resource allocation for both recovery and prevention.

Here the term "root cause" needs to be treated with care. In an accident investigation, if root cause is perceived as for example, "someone's behavior" then it may be likely, as argued by Rasmuseen (1997), that the accident would occur by another cause at another time. The author of this work agrees that, in that example, such a description of root cause is superficial and should be regarded as still part of the symptom rather than the real root cause. A real root cause needs to be plan and policy related with respect to the current status quo. As such, ideally a root cause should lead to initiation or modification of standard operating procedures (SOPs). Also a root cause needs to contribute to the three features of how learning from failures is defined as outlined by Labib and Read (2013), where it is they argued that learning from failures consists of feedback to design of existing procedures, use of advanced techniques to analyze failures, and generation of interdisciplinary generic lessons.

The innovative aspect of the study presented here is the integration of modeling approaches in a generic hybrid model. The majority of modeling applications tend to use a single model to analyze a problem, but the issue with this line of approach is that any methodology that relies on just one stand-alone, off-the-shelf model has its limitations due to the inherent assumptions that exist in any one particular model and, accordingly, it becomes inadequate in providing an effective and realistic approach. Subsequently, such an approach leads itself into trying to manipulate the problem in hand in order to "fit it" into the method instead of vice versa. So, relying on just one model for analysis may distract people who are carrying out the accident investigation, as they attempt to fit the accident into the model, which, as argued by Kletz (2001), may limit free thinking. Therefore, what is proposed here is to develop a hybrid model approach which offers richness to the analysis. Moreover, it combines the strengths of the models used and the limitations that exist in each model tend to cancel each other; in other words, they complement each other. The proposed hybrid models are fully integrated to provide an effective and efficient approach, firstly for identifying and prioritizing important features that led to the disaster and need improvement, and secondly for optimizing the allocation of resources to prevent or mitigate the consequences of future disasters.

10.2   **BACKGROUND TO THE DISASTER**

The Hurricane Katrina disaster has been studied in detail by the American Society of Civil Engineers (ASCE) (2006), Select Bipartisan Committee (2007)—a Committee of the US Senate, Jonkman et al. (2009), Crowther et al. (2007), Griffis (2007), and vanRee et al. (2011). Although it could be argued that this section could be made much shorter by simply referring the reader to the abundant information in the above reports it is suggested that, as the disaster happened a while ago, such a primary data collection would be of lower quality as memories have faded and key persons may have disappeared. Therefore, a secondary data analysis (which is a proven and widely used research method) will be used for the problem structuring. Such a secondary analysis also gives the possibility of triangulating sources. Moreover, the information can be easily checked by other researchers.

New Orleans is situated near where the Mississippi River flows into the Gulf of Mexico in south eastern Louisiana. It was built on low-lying marshland between the Mississippi River, Lake Pontchartrain, Lake Borgne, and the Gulf of Mexico. The region and its busy port, which is one of the most important in the United States, are part of Louisiana's extensive petroleum infrastructure which supplies oil and other petroleum products to the rest of the country. The state of Louisiana itself is ranked fifth in US oil production is home to a network of pipelines, storage facilities, 17 petroleum refineries, and two of the United States' four strategic petroleum reserves. Also, New Orleans serves as a business center for BP, Shell Oil, Chevron, and ConocoPhillips.

10.2.1 **A Sinking Region**

New Orleans is built on a foundation of thousands of feet of soft sand, silt, and clay. Subsidence (settling) of the ground surface occurs naturally due to consolidation, oxidation, and groundwater pumping influences. Large portions of Orleans, St. Bernard, and Jefferson parishes (counties) are therefore below sea level and continue to sink.

Prior to 1946, flooding and subsequent sediment deposition counterbalanced natural subsidence leaving south eastern Louisiana at or above sea level. However, due to major flood control structures being put in place, fresh layers of sediment are not being deposited so as to replenish ground lost to flooding. Also, groundwater withdrawal, petroleum production, development, and other factors are all contributing to the subsidence, which is estimated by the US Geological Survey to occur at a rate of between 0.15 and 0.2 in per year with rates of up to 1 in per year occurring in some places.

10.2.2 **Hurricane Protection System**

Responsibility for design and construction of most of the flood and hurricane protection levees along the Mississippi River and in the New Orleans region rests with USACE (United States Army Corps of Engineers). Their strategy was to build levees or floodwalls around segments of New Orleans. Typical USACE flood protection structures constructed in and around New Orleans were: (i) earthen levees, (ii) I-walls, and (iii) T-walls.

Table 10.1 Major Hurricanes to have Crossed Southeast Louisiana or Vicinity (1851−2005)

Hurricane	Year	Category at First Landfall	Central Pressure at First Landfall (millibars)
Camille	1969	5	909
Katrina	2005	3	920
Andrew	1992	5[a]	922
LA (New Orleans)	1915	4	931
LA (Last Island)	1856	4	934
SE FL/SE LA/MS	1947	4	940
Audrey	1957	4	945
LA (Chenier Caminanda)	1893	3	948
Betsy (SE FL/SE LA)	1965	3	948
LA/MS	1855	3	950
LA/MS/AL	1860	3	950
LA	1879	3	950
LA (Grand Isle)	1909	3	952

Source: *National Oceanic and Atmospheric Administration (NOAA) Technical Memorandum NWS TPC-4, "The Deadliest, Costliest, and Most Intense United States Tropical Cyclones from 1851 to 2005 (and other frequently requested hurricane facts)."*
[a]*Hurricane Andrew was a Category 5 hurricane in Florida but a Category 3 hurricane as it reached Louisiana.*

Also, some other agencies own and operate flood protection systems such as the interior drainage and pumping stations, the Mississippi River Levee Flood Protection System, and non-USACE levee features.

10.2.3 **Hurricane Katrina**

Hurricanes are not a new phenomenon to southeast Louisianans as can be seen from Table 10.1.

Hurricane Katrina started out in the Bahamas, as a tropical storm, on August 23, 2005. It crossed south Florida on 25 August as a Category 1 hurricane before entering the Gulf of Mexico, intensifying as it tracked westward.

Hurricanes are categorized according to their maximum wind speed, on the Saffir−Simpson Hurricane Scale, as illustrated in Table 10.2.

Table 10.2 The Saffir–Simpson Hurricane Scale

Category	Wind Speed (mph)	Typical Storm Water Surge (ft)
1	74–95	4–5
2	96–110	6–8
3	111–130	9–12
4	131–155	13–18
5	>155	>18

On 28 August, the hurricane began tracking toward the northwest and intensified from a Category 2 to a Category 5 in just 12 h. As it approached land, the warm, moist air, and energy that it could draw from the Gulf of Mexico decreased, and Hurricane Katrina was degraded to a Category 3.

On August 29, 2005 the hurricane, which was one of the strongest storms ever to hit the coast of the United States, brought intense winds, high rainfall, waves, and storm surges that caused widespread devastation in New Orleans and along the coasts of Louisiana, Mississippi, and Alabama. Levees and floodwalls were overtopped and several were breached, allowing billions of gallons of water from the Gulf of Mexico, Lake Borgne, and Lake Pontchartrain to flow into New Orleans, flooding major portions of the city.

10.2.4 **Consequences of Failure**

The following statistics are attributed to the Katrina disaster:

- *Fatalities*: As of August 2, 2006, 1118 people confirmed dead
- *Damage to residential and nonresidential property*: $21 billion
- *Damage to public infrastructure*: $6.7 billion
- *Population displaced*: Approximately 50%
- *Regional economy*: Approximately 124,000 jobs lost and region's economy crippled

10.3 **TECHNICAL CAUSES OF FAILURE**

From the many causes identified in previous studies (ASCE, 2007; Crowther et al., 2007; Select Bipartisan Committee, 2007; Jonkman et al., 2009) as being responsible for the disaster the high-level ones can be categorized as either direct or contributory.

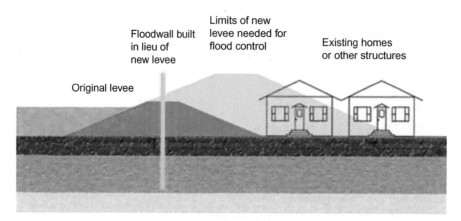

■ **Figure 10.2** Increased levee elevation using I-wall in lieu of new levee.

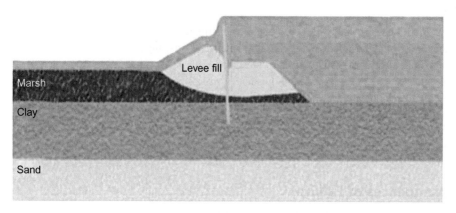

■ **Figure 10.3** Soft I-wall foundations resulting in overtopping and collapse.

10.3.1 **Direct Causes**

a. *Levees breached*: Authorities used I-walls to raise levee elevations instead of increasing width and height with earth, wrong elevation datums were used and they did not account for soft I-wall foundations, resulting in overtopping and I-wall collapse. Also many infrastructure penetrations were made through levees (Figures 10.2 and 10.3).

b. *Ineffective pumping stations*: Underrated pumping stations which were not hurricane resistant required the presence of an operator and power to function, resulting in only 16% utilization of the limited pumping capacity for the region.

10.3.2 **Contributing Factors**

c. *Hurricane protection management policy*: Poor risk management approach, uncoordinated construction, maintenance, and operation of levees, also no levee subsidence correction program, all contributed to the unfolding of the disaster.

d. *Inadequate emergency response*: Lack of a mandatory evacuation order, unprepared local and state emergency response agencies and a late national response impeded rescue efforts.

10.4 **HYBRID MODELING APPROACH**

As mentioned earlier, the innovative aspect of this study is the integration of modeling approaches in a generic hybrid model. In order to show how the different tools and techniques are coordinated, Table 10.3 briefly explains each one used in terms of the sources of data used, the outcome of each one, and how it integrates with other tools.

The integration of the tools listed in Table 10.3 will be further illustrated in Figure 10.14. The AHP was used rather than the analytic network process (ANP), although the latter allows interaction between the different criteria because AHP is hierarchical in nature which is in line with the architecture of FTA. Furthermore, the simplicity in computation of AHP compared to ANP justifies its use, especially in a situation where hybrid modeling is used.

10.5 **FAULT TREE ANALYSIS**

The fault tree shown in Figure 10.4 shows how the component failures of the direct causes of the Katrina disaster, and the contributing factor, interacted with each other. Its construction was guided by technical causes of the failure and understanding of the problem as derived from the sources listed in Table 10.3.

Starting at the top event, or system failure (the Hurricane Katrina disaster), the fault tree was built downward from the undesired event to intermediate events (rectangular boxes) and then to basic events using logical AND/OR gates and deductive logic, i.e., repeatedly asking "what are the reasons for this event?" The basic events of the fault tree refer to the component failures and human errors and are indicated in the circular/elliptical shapes at the bottom of the diagram.

The AND/OR gates describe the fault logic between the events, i.e., the logic gate underneath the intermediate event "Levees breached" is an

Table 10.3 Integration of Tools with the Proposed Hybrid Model

Tools	Data Used by Each Tool	Tool's Usage and Outputs	Figures and Tables That Show Results of the Tool
FTA	Needs data related to technical causes of the failures, based on understanding of the problem and on previously published research (e.g., ASCE, 2007; Crowther et al., 2007; Griffis, 2007; Select Bipartisan Committee, 2007; Jonkman et al., 2009; vanRee et al., 2011)	Provides information on failure modes and their root causes in the form of basic events to FMECA tool. Provides information on interdependence in the form of OR and AND gates to RBD tool. Provides hierarchical model to AHP tool.	Figures 10.4 and 10.6
RBD	Needs basic events from the FTA tool.	Provides information in terms of "series" and "parallel" structures to inform about the relative weights assigned in the AHP tool.	Figure 10.5
	Needs relationship between basic event from FTA, i.e., every OR gate in FTA is mapped as a "series" structure in RBD, and every AND gate in FTA is a "parallel" structure in RBD.	Demonstrates how the overall system reliability can be calculated and improved by strengthening weak (series) structures in the diagram.	
FMECA	Needs information on failure modes from the FTA tool.	Provides information on RPN.	Tables 10.7−10.10
AHP	Needs information on RPN from FMECA tool.	Provides a hierarchical structure of the problem from FTA tool.	Figure 10.8(a) and (b)
	Needs information about hierarchical model from FTA tool.	Provides prioritization for selection or resource allocation.	Figures 10.10 and 10.11
	Needs information in terms of series and parallel structures from RBD tool.	Provides sensitivity analysis— "what−if" analysis.	Figures 10.12 and 10.13
	Needs information about judgments from questionnaire completed by decision makers and informed by structure of FTA and RBD tools.	Provides a consistency feedback measure to the decision maker.	Figures 10.9 and 10.11

"OR" gate. The inputs to this gate are "I-wall collapse," "overtopping," and "infrastructure breaches." Connecting these events using OR logic shows that this means that if any of these intermediate events occur, then they will cause the output event to occur, i.e., "Levees breached." If the type of gate used was instead an AND gate, then this would indicate a level of redundancy in the system, in that all three basic events/component failures would have to occur before the output event would occur. For both AND and OR gates, the minimum number of inputs required to make these gates valid is two.

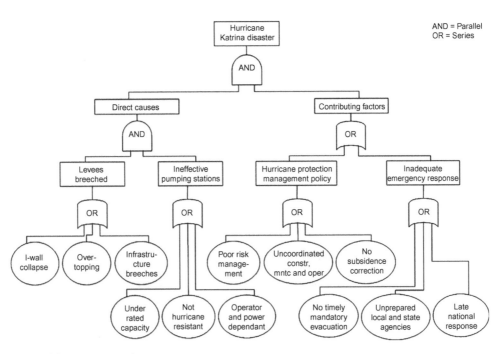

■ **Figure 10.4** FTA of the Hurricane Katrina disaster.

Symbols for OR and AND gates are shown in Table 10.4.

For AND and OR gates, the probabilities are calculated as follows:

$$P_{(AND\_Gate)} = \Pi P_{(input\_i)}$$

$$P_{(OR\_Gate)} = 1 - \Pi(1 - P_{(input\_i)})$$

For AND gate, the following logic applies as shown in Table 10.5.

For OR gate, the following logic applies as shown in Table 10.6.

An FTA can be used to achieve the following:

- Define the undesired event to study.
- Obtain an understanding of the system.
- It provides a useful graphical representation of the hierarchy of failure modes, a helpful mental map of the logic.
- Diagnostics and fault finding, because it follows a logical and systematic process of breaking down a complex problem into its root causes.

Table 10.4 OR and AND Logic Gates

Symbol	Symbol Name	Description	Reliability Model	Inputs
	OR gate	The output event occurs if any of its input event occur	Failure occurs if any of the parts of the system fails—series system	≥ 2
	AND gate	The output event occurs if all input events occur	Parallel redundancy, one out of n equal or different branches	≥ 2

Table 10.5 AND Gate Logic

$P_{input\ 1}$	$P_{input\ 2}$	P_{output}
0	0	0
1	0	0
0	1	0
1	1	1

Table 10.6 OR Gate Logic

$P_{input\ 1}$	$P_{input\ 2}$	P_{output}
0	0	0
1	0	1
0	1	1
1	1	1

- Quantification, although it is inherently a qualitative tool. It allows qualitative analysis to be performed, i.e., the component failure combinations which result in the top event or system failure occurring. But it also allow quantitative analysis to be performed, i.e., of the probability of system failure, frequency of failure, system downtime, and expected number of failures in x years.
- Attention focus on the top event, a certain critical failure mode, or an event (which can be deductively cascaded).

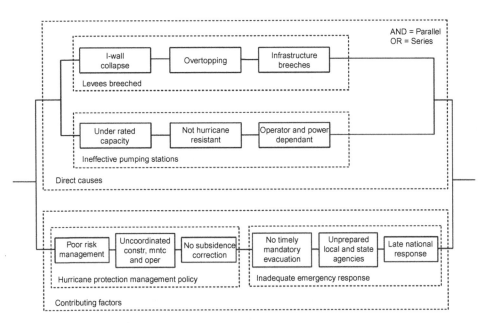

■ **Figure 10.5** RBD of the Hurricane Katrina disaster.

- Identification of the items of the system which are important to the particular failure being studied.
- Analysis of human, software, procedural, and other factors, in addition to the role of the physical parts of the system.

10.6 **RELIABILITY BLOCK DIAGRAM**

The RBD of the disaster, as shown in Figure 10.5, gives added value to the analysis, as it provides the decision maker with a better understanding of the overall reliability of the model by highlighting vulnerable aspects, where series dependencies exist and relatively safe areas where there is either redundancy or parallel dependency. Moreover, given reliability values for the different boxes (components), one can calculate the whole system's reliability. So, in order to maximize system reliability and minimize system failure rate, then the number of series dependencies (components in series) should be kept to a minimum. RBDs can be used as follows:

- They allow quantitative analysis to be performed.
- The interconnections within the RBD symbolize the way in which the system functions as required and in which that is determined by the reliability dependencies.

- It can usually be constructed on the basis of information provided within an FS (functional specification) document, a schematic or P&ID (piping and instrumentation drawing).
- It can be constructed using a two-state assumption, i.e., either fully operational (up) or totally failed (down), which renders subsequent analysis much easier than it otherwise would be if a third state existed, i.e., partially failed (e.g., reduced output from a pumping station).
- The two-state assumption, which is conservative, makes reliability evaluations desirably cautious when safety assessments are involved.

However, it has been argued by some critics that such analysis cannot be realistically quantified and that results are irrelevant when there is a high degree of subjectivity. It should be emphasized here that in this study of Katrina event, the incorporation of both FTA and RBD methods is intended for a *risk-informed* rather than a *risk-based* decision making (as outlined by Apostolakis, 2004). Hence the approach proposed here is intended to address the concerns of both engineers and designers, in an effort to facilitate the mental modeling process for understanding the nature of factors contributing to disasters, rather than satisfying an accountant's objective of determining an accurate figure of the cost of risk or safety issues. So we aim to understand failure modes rather than quantify probabilities per se. This is analogous to the fuzzy logic concept where it attempts to combine subjective human judgments and rules—the "fuzzy" part—with the mathematical rigor of operational research techniques to achieve rational decision making.

Apostolakis (2004) provided a balanced approach with respect to the benefits and limitations of such techniques and argued that while they are not perfect they represent a considerable advance in rational decision making. Our approach is in line with the three questions posed by Kaplan and Garrick (1981), namely: (1) What can go wrong? (2) How likely is it? and (3) What are the consequences? So our approach in using reliability analysis is stronger on relativities (i.e., relative comparisons and trends) but weaker on absolutes. Nevertheless, it is believed that a candle in the dark is better than no light at all.

10.7　**FAILURE MODES, EFFECTS, AND CRITICALITY ANALYSIS**

As regards only the direct causes of the Katrina disaster, a basic failure modes, effects, and criticality analysis template, as given Table 10.7, can be used to document for each component mode of failure its effects

Table 10.7 FMECA of Direct Causes of Hurricane Katrina

Item No.	Component and Mode of Failure	Effect, i.e., Symptoms	Cause of Failure	Probability of Occurrence of Failure (O)	Severity (S)	Difficulty of Detection (D)	Risk Number (RPN)	How Can the Failure Be Eliminated or its Effect Reduced
1	Levee: I-wall collapse	Immediate flooding of protected areas	Low safety margin used based on meteorological conditions for area, i.e., maximum Category 3 barometric pressure and wind speed assumed (inherent effects of reduction from category 4/5 to 3 not factored in)	3	5	3	45	1. Strengthen/ replace I-walls 2. Segregate areas by use of internal levees to contain breaches
2	Levee: Overtopping	Initial slow flooding of protected areas, followed by fast flooding erosion and destruction of foundations	Soft soils beneath and adjacent to levees unprotected	3	4	4	48	1. Harden soils beneath/ adjacent to levees by mixing in/replacing with aggregate 2. Protect soils adjacent to levees with concrete
3	Levee: Overtopping	Fast flooding of protected areas	Incorrect design elevations (1–2 ft) due to use of wrong datum	3	5	2	30	Raise levees to correct elevation

Continued

Table 10.7 FMECA of Direct Causes of Hurricane Katrina *Continued*

Item No.	Component and Mode of Failure	Effect, i.e., Symptoms	Cause of Failure	Probability of Occurrence of Failure (O)	Severity (S)	Difficulty of Detection (D)	Risk Number (RPN)	How Can the Failure Be Eliminated or its Effect Reduced
4	Levee: Infrastructure Breaches	Initial slow flooding of protected areas, followed by fast flooding on further erosion of breach	Many penetrations through levees for roads, railroads, and utilities	4	4	2	32	Eliminate/ redesign levee penetrations
5	Pumping stations: underrated capacity	Floodwaters in protected areas continue to rise	Underrated, i.e., only suitable for storm water runoff and routine seepage of water from interior drainage system	3	4	4	48	Increase pumping capacity to cope with storm surges from surrounding bodies of water
6	Pumping stations: location	Floodwaters remain in protected areas	Stations not located in areas worst hit by floods	3	4	3	36	Build pumping stations in areas likely to flood
7	Pumping stations: piping design	Floodwater recirculation back to body of water where it came from	Discharges hard piped back into canals and waterways	4	4	2	32	Reroute discharges away from city and bodies of water so that recirculation cannot occur

8	Pumping stations: backflow	Water backflowed to city through inoperable pumps/discharge pipes	3	No automatic backflow preventers in place	4	2	24	Install automatic backflow preventers
9	Pumping stations: building design	Pumping stations became inoperable	3	Not designed to withstand hurricane forces and hence damaged by hurricane	3	4	36	Reinforce/ rebuild so as to withstand hurricane forces
10	Pumping stations: operation	Pumping stations became inoperable	3	Dependant on operators who were evacuated and no automatic control	3	3	27	Automate pumping stations so that they can operate without operators present
11	Pumping stations: power	Pumping stations became inoperable	4	Dependant on electricity which was knocked out early on and no backup generation	3	2	24	Install redundancy in power supply lines and onsite backup generation

(symptoms), the cause of that failure and how the failure can be eliminated or its effect reduced. For example, Item No. 1 of the template would be populated as follows:

> *Component and mode of failure* = Levee: I-wall collapse.
> *Effect, i.e., symptoms* = Immediate flooding of protected areas.
> *Cause of failure* = Low safety margin used based on meteorological conditions for area, i.e., maximum Category 3 barometric pressure and wind speed assumed (inherent effects of reduction from Category 4/5 to 3 not factored in).
> *How can the failure be eliminated or effect reduced* = 1. Strengthen/replace I-walls and 2. Segregate areas by use of internal levees to contain breaches.

Next, by constructing a word model as per Table 10.7, and applying it to the template, it is possible to rank each component mode of failure in terms of its probability of occurrence, severity, and difficulty of detection so as to arrive at an overall RPN which can then be used to focus attention on the highest risk items. The information in Table 10.7 and its numerical probabilities were derived from the author's understanding of the published literature on the Katrina disaster as outlined previously.

Tables 10.8–10.10 are then used in deriving the figures shown in Table 10.6. The RPNs have no units, and the objective of the FMECA is to reduce the RPNs; they can be calculated as follows:

$$RPN = \textit{Probability of occurrence of failure} \times \textit{Severity} \\ \times \textit{Difficulty of Detection } (O \times S \times D).$$

Table 10.8 RPN Word Model for FMECA of Direct causes of Hurricane Katrina, in Terms of Probability of Occurrence of Failure

Probability of Occurrence of Failure (O)		
Keyword	**Time Interval**	**Score**
Very unlikely	>50 years	1
Unlikely	20–50 years	2
Possible	10–20 years	3
Probable	5–10 years	4
Frequent	<5 years	5

Table 10.9 RPN Word Model for FMECA of Direct Causes of Hurricane Katrina, in Terms of Severity

Severity (S)	
Keyword	**Score**
Repair cost only	1
Minor property damage	2
Significant property and minor economy damage	3
Major property and significant economy damage	4
Loss of life, major property and economy damage	5

Table 10.10 RPN Word Model for FMECA of Direct Causes of Hurricane Katrina, in Terms of Difficulty of Detection

Difficulty of Detection (D)	
Keyword	**Score**
Almost certain	1
High	2
Moderate	3
Low	4
Absolute uncertainty	5

Thus, with Item No. 1 of the example, the rankings would be populated and the RPN calculated as follows:

Probability of occurrence of failure (O) = 3, i.e., possible in time interval 10–20 years.
Severity (S) = 5, i.e., loss of life, loss of major property, and damage to the economy.
Difficulty of detection (D) = 3, i.e., moderate.

So, $RPN = 3 \times 5 \times 3 = 45$

An FMECA can be used to achieve the following:

- Systematic determination of the ways in which failure can occur and the effects (from minor to catastrophic) that each such failure can have on overall functionality.
- Anticipation of failures and identification of the means of their prevention.
- Prioritization of the failure modes for corrective action.

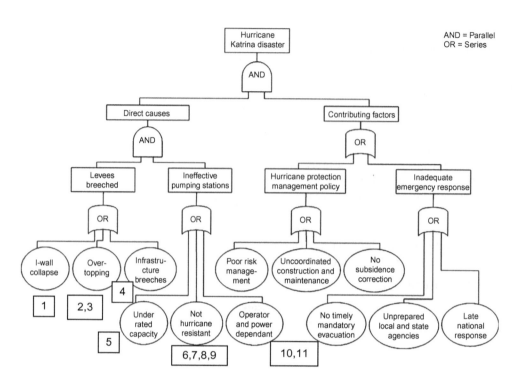

■ **Figure 10.6** FTA of the Hurricane Katrina disaster with item numbers given in Table 10.7.

Bradley and Guerrero (2011) have acknowledged the wide use of FMECA in both product design and industry. However, they have identified two prominent criticisms of its traditional application, namely that according to measurement theory, RPN is not a valid measure for ranking failure modes, and that it does not weight the three decision criteria used in FMECA. They have proposed a new ranking method that can overcome this criticism. Here, however, we use traditional FMECA and then use AHP to deal with the ranking issues.

The item numbers in Table 10.7 of the FMECA correspond to the items in the FTA in Figure 10.4 and as shown in Figure 10.6.

10.8 AN AHP MODEL FOR THE HURRICANE KATRINA DISASTER

The first step in the AHP approach as described in Chapter 3 is the decomposition of the problem into a decision hierarchy. This may take the form illustrated in Figure 10.7.

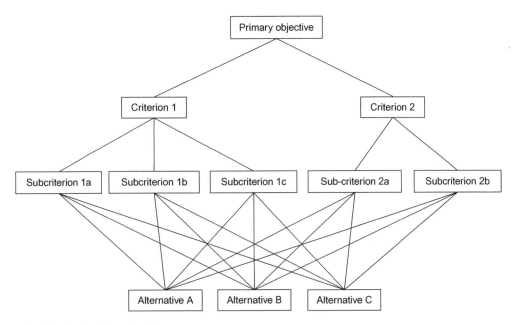

■ **Figure 10.7** A hierarchical model based on AHP.

The FTA model shown in Figure 10.4 is used as the hierarchical model in an AHP analysis, where the higher levels are criteria and subcriteria, and the basic events (input values) are considered as alternatives using the terminology employed in Figure 10.6. The diagram shown in Figure 10.4 is then translated into an AHP hierarchy, as shown in Figure 10.8(a) and (b), where the left side shows criteria and subcriteria, and the right side the different alternatives (basic events).

Pairwise comparison is then performed on each level with respect to the level above. The pairwise judgment being informed by the FTA model of Figure 10.4. For example, when comparing *"Direct Causes"* to *"Contributing Factors"* more relative weight is given to *"Contributing Factors"* than to *"Direct Causes"* as shown in Figure 10.8. The rationale behind this is that interdependence is examined of the events below those two factors. Under *"Direct Causes,"* there is an AND gate, which implies that there are two factors that have *simultaneously* affected the direct causes, namely *"Levees Breached"* AND *"Ineffective Pumping Stations."* This implies an element of redundancy. Whereas under *"Contributing Factors,"* there is an OR gate, implying that either *"Hurricane Protection Management Policy"* OR *"Inadequate Emergency Response"* can have a direct effect on the *"Contributing Factors."* This is also clearly

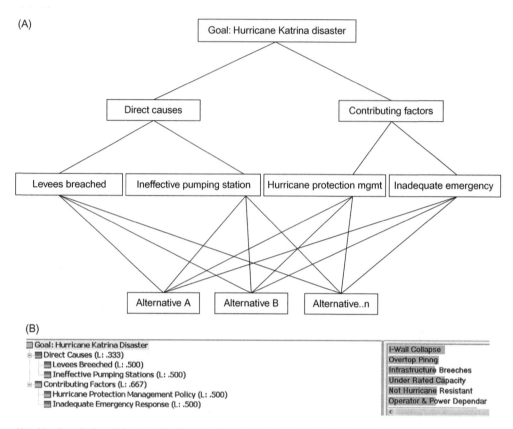

■ **Figure 10.8** (a) A hierarchical model based on AHP, (b) A hierarchical model based on AHP using Expert Choice software.

demonstrated in the RBD model of Figure 10.5, where the *"Direct Causes"* are modeled in a relatively safer parallel structure as compared to the *"Contributing Factors,"* which are modeled in series structure, which is less vulnerable (Aven, 2011).

The pairwise judgment is also informed by Table 10.7 FMECA model. However, FMECA is only applicable for failure modes and hence RPN values are only provided for items under *"Direct Causes."* Items under *"Contributing Factors"* are subjectively assessed. The RPN given in Table 10.7 provides useful information about the rankings of the different risks (failure modes) associated with the Katrina disaster.

A questionnaire can be used to perform the pairwise comparison, as shown in Figure 10.9. This employs Saaty's scale, as given in Table 10.11.

Compare the relative preference

I-WALL COLLAPSE	*versus*	OVERTOP PINNG

with respect to: Levees Breeched (L: .500)

		9 8 7 6 5 4 3 [2] 1 2 3 4 5 6 7 8 9	
1	I-Wall Collapse	9 8 7 6 5 4 3 [2] 1 2 3 4 5 6 7 8 9	Overtop Pinng
2	I-Wall Collapse	9 8 7 6 5 4 [3] 2 1 2 3 4 5 6 7 8 9	Infrastructure Breeches
3	I-Wall Collapse	[9] 8 7 6 5 4 3 2 1 2 3 4 5 6 7 8 9	Under Rated Capacity
4	I-Wall Collapse	[9] 8 7 6 5 4 3 2 1 2 3 4 5 6 7 8 9	Not Hurricane Resistant
5	I-Wall Collapse	[9] 8 7 6 5 4 3 2 1 2 3 4 5 6 7 8 9	Operator & Power Dependant
6	I-Wall Collapse	[9] 8 7 6 5 4 3 2 1 2 3 4 5 6 7 8 9	Poor Risk Management
7	I-Wall Collapse	[9] 8 7 6 5 4 3 2 1 2 3 4 5 6 7 8 9	Uncoordinated Construction, N
8	I-Wall Collapse	[9] 8 7 6 5 4 3 2 1 2 3 4 5 6 7 8 9	No Subsidence Correction
9	I-Wall Collapse	[9] 8 7 6 5 4 3 2 1 2 3 4 5 6 7 8 9	No Timely Mandatory Evacuati
10	I-Wall Collapse	[9] 8 7 6 5 4 3 2 1 2 3 4 5 6 7 8 9	Unprepared Local & State Age
11	I-Wall Collapse	[9] 8 7 6 5 4 3 2 1 2 3 4 5 6 7 8 9	Late National Response
12	Overtop Pinng	9 8 7 6 5 4 3 [2] 1 2 3 4 5 6 7 8 9	Infrastructure Breeches
13	Overtop Pinng	[9] 8 7 6 5 4 3 2 1 2 3 4 5 6 7 8 9	Under Rated Capacity
14	Overtop Pinng	[9] 8 7 6 5 4 3 2 1 2 3 4 5 6 7 8 9	Not Hurricane Resistant

1 = Equal	3 = Moderate	5 = Strong	7 = Very Strong	9 = Extreme

■ **Figure 10.9** A questionnaire of judgments based on pairwise comparisons.

Table 10.11 Saaty's Rating Scale

Intensity of Importance	Definition	Description
1	Equal importance	Two factors contribute equally to the objective.
3	Somewhat more important.	Experience and judgment slightly favor one over the other.
5	Much more important.	Experience and judgment strongly favor one over the other.
7	Very much more important	Experience and judgment very strongly favor one over the other. Its importance is demonstrated in practice.
9	Absolutely more important	The evidence favoring one over the other is of the highest possible validity.
2,4,6,8	Intermediate values.	When compromise is needed

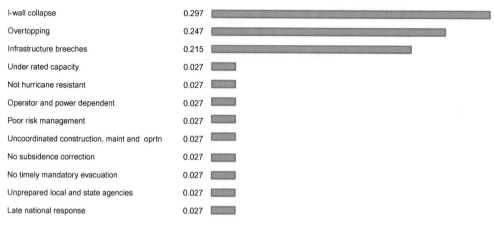

Priorities with respect to Goal: Hurricane Katrina disaster > Direct causes > Levees breeched

I-wall collapse	0.297
Overtopping	0.247
Infrastructure breeches	0.215
Under rated capacity	0.027
Not hurricane resistant	0.027
Operator and power dependent	0.027
Poor risk management	0.027
Uncoordinated construction, maint and oprtn	0.027
No subsidence correction	0.027
No timely mandatory evacuation	0.027
Unprepared local and state agencies	0.027
Late national response	0.027

Inconsistency = 0.00851

■ **Figure 10.10** Priorities of items under *"Levees Breached."*

Note that in Figure 10.9, the questionnaire relates to the judgments with respect to failure mode *"Levees Breached."* We know from the FTA in Figure 10.6 that under *"Levees Breached,"* the three items of *"I-Wall Collapse," "Levee Overtopping,"* and *"Infrastructure Breaches"* are the most relevant and hence take the highest weights, of value of 9 (extreme), compared to the rest of the items. We also know from the RPN values in Table 10.8 that *"I-Wall Collapse"* ranks slightly higher than *"Levee Overtopping"* which is in turn slightly higher than *"Infrastructure Breaches."* This sort of logic for elicitation of judgment is shown in Figure 10.10.

The same comparison is then carried out across the rest of the hierarchy. Once the pairwise comparison is completed, the calculation of the overall ranking of the basic events is obtained, as shown in Figure 10.11.

There are three outputs that can be produced from this AHP model, namely:

1. Overall rankings as shown in Figure 10.11, where the summation of the rankings is equal to unity. This aids the allocation of resources among alternative basic events. It also helps in understanding how each basic event can be compared to the others.
2. A measure of overall inconsistency of the decision maker's preferences, which is a useful feedback for validation of consistency. Overall inconsistency less than 10% is normally acceptable as a

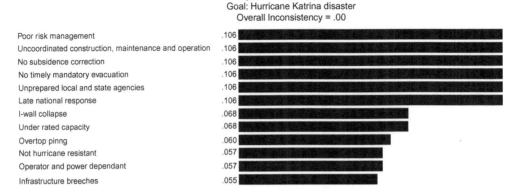

Synthesis with respect to:
Goal: Hurricane Katrina disaster
Overall Inconsistency = .00

Poor risk management	.106
Uncoordinated construction, maintenance and operation	.106
No subsidence correction	.106
No timely mandatory evacuation	.106
Unprepared local and state agencies	.106
Late national response	.106
I-wall collapse	.068
Under rated capacity	.068
Overtop pinng	.060
Not hurricane resistant	.057
Operator and power dependant	.057
Infrastructure breeches	.055

■ **Figure 10.11** Global priorities of alternatives (basic events).

measure of consistency of preferences. Saaty (1980) has developed a randomly generated matrix for each matrix dimension, which is called "consistency index" (CI). Essentially, consistency is measured by providing a degree of closeness to a randomly generated matrix of judgments in the form of a consistency ratio (CR). So the further away from randomness, the more consistent are the preferences.

3. A facility to perform sensitivity analysis (what−if analysis) provides information about the causal relationships between the different factors. This capability can help us to explain and predict the different relationships between criteria and alternatives. The sensitivity analysis is illustrated in Section 10.9.

It is worth noting here that pairwise comparisons between failure mechanisms could have been done by comparing limit state functions of strength and loading variables, instead of subjective scoring on a scale of 1−9. However, it is argued that the ability to handle subjectivity is the strength of the AHP method. Moreover, AHP offers both sensitivity analysis and feedback on consistency.

10.9 **RESULTS OF SENSITIVITY ANALYSIS**

The results of the sensitivity analysis are shown in Figure 10.12, where the criteria are depicted as columns; e.g., the criterion of "*Contributing Factors*" is more important than the criterion "*Direct Causes*" as explained before. It also shows the performance of each of the alternatives (basic events) with respect to each of the criteria. For example, with respect to "*Direct Causes*" (at the first column from the left), we note

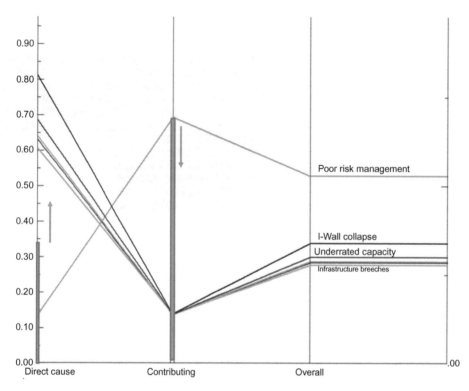

■ **Figure 10.12** Sensitivity analysis.

that basic event "*I-Wall Collapse*" ranks higher (more significant) than "*Overtopping*," whereas with respect to "*Contributing Factors*," we note that "*Poor Risk Management*" ranks higher (more significant) than the rest. Also, taking into consideration all the different criteria, the overall ranking of basic events is shown at the column titled "Overall."

Now, what happens if the importance of the criterion "*Direct Causes*" is increased and that of "*Contributing Factors*" decreased? This is illustrated by the directions of both arrows in Figure 10.12. We note in Figure 10.13 that the overall ranking of alternative basic events (as shown on the right side of the figure) will change into "*I-Wall Collapse*" higher than "*Underrated Capacity*," which in turn is higher than "*Overtopping*," and so on. Hence the overall rankings of alternatives change according to the change in the weights of controlling criteria. This "what–if" analysis is a very powerful facility, as it can help us to predict the importance of criteria in changing environments that will subsequently affect the importance of different alternatives.

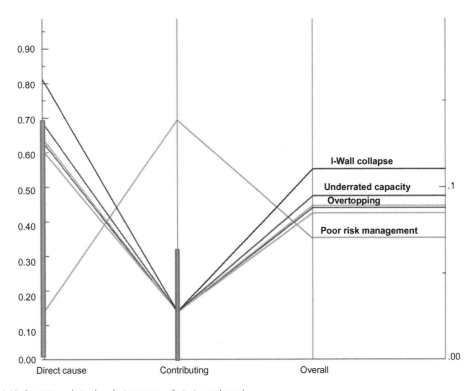

■ **Figure 10.13** Sensitivity analysis when the importances of criteria are changed.

This analysis provides a powerful tool for enhancing our understanding of how different factors interact with, and affect, other factors.

10.10 **DISCUSSION AND LESSONS LEARNED**

In addition to the actions identified in Sections 10.5, 10.6, and 10.8, which should be implemented so as to eliminate or reduce the possibility of failure of the hurricane protection system, the following high-level design improvements/lessons learned were also identified for action.

1. Assign single entity responsibility for managing critical hurricane and flood protection systems, ensuring:
   a. The entity is organized and operated to enable, not inhibit, a focus on public safety, health, and welfare.
   b. Public safety, health, and welfare are the top priorities.
2. Establish a mechanism for a nationwide levee safety program, similar to that in place for dams.

3. Quantify and periodically update the assessment of risk.
4. Determine the level of acceptable risk in communities through quality interactive public risk communication programs and manage risk accordingly.
5. Establish continuous engineering evaluation of design criteria appropriateness, always considering the impact of individual components on the overall system.
6. Correct hurricane and flood protection system physical deficiencies by establishing mechanisms to incorporate changing information.
7. Implement more effective mechanisms to ensure coordination and cooperation between designers, maintainers, and operators, e.g., apply concurrent engineering techniques.
8. Upgrade engineering and design procedures and practices to place greater emphasis on safety.
9. Engage independent experts in high-level review of hurricane and flood protection systems.

Figure 10.14 shows how the different techniques employed are linked into a cohesive and integrated model of the kind applied here to an analysis of the Hurricane Katrina disaster.

10.11 **CONCLUDING REMARKS**

The main contribution of this analysis of the causes of the Katrina disaster is that it shows that the proposed methodology can produce better information for policy makers. This is achieved through the utilization of selective operational research and reliability analysis related techniques in an integrated approach. Such methodology supports modeling of the factors that lead to a disaster and then provides a facility for taking rational decisions. So this prompts the question: What is new in the proposed theory? Is it novel relationships? Or is it better decisions? It is probably both of these, but with emphasis on mental modeling, as it is believed that formulating a problem normally solves 80% of it. This is in line with the argument posed by Einestein and Infeld (1938) "The *formulation of a problem is often more essential than its solution, which may be merely a matter of mathematical or experimental skill*." It is acknowledged that every technique has its own limitation, and hence the originality of the approach lies in utilizing a hybrid of techniques that tend to cancel the limitations inherent in any one of the used techniques when used on its own.

In terms of reliability analysis, FTA has been employed to show how some of the direct causes and contributing factors of the disaster interacted with each other. It provided information on the failure modes that

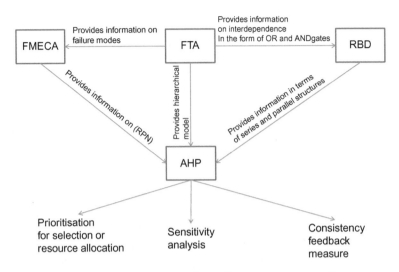

■ **Figure 10.14** The relationships between the different techniques in the integrated model.

have been used in the FMECA model. Also, it provided information on interdependence, in the form of OR and AND gates, which was useful for the RBD model and finally it provided a hierarchical structure, which improved our understanding of the problem, and that formed the basis for the developed AHP model. RBD representation has then been used to demonstrate how overall system reliability can be calculated and improved. RBD provided information, in terms of series and parallel structures, which has been used to inform our elicitation of judgments in the AHP model. The FMECA model has then been used to determine RPN which has been used to rank the risk of different failure modes. Subsequently, this has informed our judgments in the pairwise comparison within the AHP model.

In terms of MCDM, we have utilized an AHP analysis which was able to provide prioritization, sensitivity analysis, and feedback on consistency of the different criteria and alternative contributing factors. Finally, some of the high-level design improvements/lessons learned, which should be acted upon so as to avoid a repeat disaster, have been discussed.

A generic approach has been offered that can be used for risk and safety analysis in order to learning from disasters. This approach can also be applied to previous research studies of process and nuclear disasters such as those of Pate-Cornell (1993), and Vaurio (1984).

It can be argued that the fault tree in Figure 10.5 is as a more general logic tree rather than a strict classical fault tree as traditionally used to analyze failures at equipment level. The reason is that the "Hurricane Katrina disaster" is depicted as a top event (which is not strictly traditional in the usual fault tree sense, i.e., it is not a well-defined subset of a sample space, to which a binary indicator variable can be attached—1 for occurrence, 0 for nonoccurrence). The advantage of using the proposed approach is that it offers richness to the model, so that both subjective judgment and objective evaluation measures are taken into account as a mental model.

Although this work has indicated how some current ranking and sensitivity analysis methods might be applied to the example of the Katrina disaster, more research is needed on how (or whether) doing so would improve risk management decisions more than other approaches. However, it is believed that the approach presented can significantly improve the objectivity and of risk management decisions.

Chapter 11

NASA's Space Shuttle Columbia Accident

11.1 WHAT HAPPENED?

The loss of Columbia was caused by damage sustained during launch, when a piece of foam insulation the size of a small briefcase broke off the main propellant tank under the aerodynamic forces of the launch. The debris struck the leading edge of the left wing of the Number 8 reinforced—carbon—carbon (RCC) tile, damaging the shuttle's thermal protection system (TPS).

The space shuttle was designed as an airplane-like orbiter with two solid rocket boosters (SRBs) on either side, and a large cylindrical external tank holding cryogenic fuel for the orbiter's main engines (Smith, 2003). The Space Transportation System (STS), designated STS-107, was the space shuttle program's 113th flight and Columbia's 28th. Columbia was the first space-rated orbiter, and it made the space shuttle program's first four orbital test flights. Unlike orbiters Challenger, Discovery, Atlantis, and Endeavor, Columbia's payload was insufficient to make it cost-effective for space station missions. Columbia was therefore not equipped with a space station docking system. Consequently, it generally flew science missions and serviced the Hubble space telescope (Starbuck and Farjoun, 2005).

The National Aeronautics and Space Administration (NASA) launched Columbia on its STS-107 mission on January 16, 2003. On February 1, as it descended to Earth after completing a 16-day scientific research mission, it broke apart over northeastern Texas. The disaster occurred at an altitude of 203,000 ft and the shuttle was moving at 12,500 miles/h. Solar wind was above 800 km/s (in the red zone), but average solar wind speed should be around 300—400 km/s. Communication was lost at 9.00 am EST. At 206,000 ft data was lost from the temperature sensors on the left wing and then from the tire pressure sensors on the left main landing gear. The left wing began to break up and tanks of hydrogen, oxygen, and propellants began to explode. Columbia's previous 27 missions into space had been successful.

Learning from Failures. DOI: http://dx.doi.org/10.1016/B978-0-12-416727-8.00011-4

11.2 **LOGIC OF THE TECHNICAL CAUSES OF THE FAILURE**

According to Smith (2003), NASA identified three broad categories of causes of the accident.

1. Physical

   The physical cause was damage to Columbia's left wing by a 1.7 pound piece of insulating foam that detached from the left "bipod ramp," which connected the external tank to the orbiter and struck the orbiter's left wing 81.9 s after launch. As Columbia descended from space into the atmosphere, the heat produced by the air molecules colliding with it typically caused wing leading-edge temperatures to rise steadily and reach an estimated 2500°F (1400°C) during the next 6 min. The foam strike created a hole in an RCC panel on the leading edge of the wing, allowing the hot air to enter the wing during reentry. The extreme heat caused the wing to fail structurally, creating aerodynamic forces that led to the disintegration of the orbiter.

2. Organizational

   The accident was rooted in the space shuttle program's history and culture, which included the original compromises that were required to gain approval for the program, the subsequent years of resource constraints, fluctuating priorities, schedule pressures, mischaracterization of the shuttle as operational rather than developmental, and lack of an agreed national vision for human space flight. Also, the crew of the Columbia mission were not informed about the launch damage.

3. "Broken safety culture" at NASA

   Schedule pressure related to construction of the International Space Station, budget constraints, and workforce reductions were also factors. The shuttle program had operated in a challenging and often turbulent environment.

11.3 **CONSEQUENCES AND SEVERITY**

This mission claimed seven crucial lives of elite NASA astronauts. Following the loss of Columbia, the space shuttle program was suspended. A huge financial loss was estimated of around 1.40 billion USD. A helicopter crashed while searching for debris, claiming a further five lives (Figure 11.1).

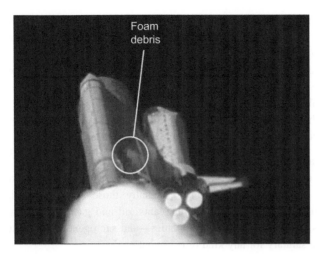

■ **Figure 11.1** Columbia space shuttle. *Courtesy NASA. NASA (page 34 of the CAIB's report (Volume 1)).*

11.4 **PROPOSED IMPROVEMENTS AND GENERIC LESSONS**

Within 2 h of losing the signal from the returning spacecraft, NASA's Administrator established the Columbia Accident Investigation Board (CAIB) to uncover the conditions that had produced the disaster and to draw inferences that would help the US space program to emerge stronger than before (CAIB, 2003). Seven months later, the CAIB released a detailed report that included its recommendations (Starbuck and Farjoun, 2005).

The CAIB (2003) report attempted to seek answers to the following four crucial questions:

1. Why did NASA continue to launch spacecraft despite many years of known foam debris problems?
2. Why did NASA managers conclude, despite the concerns of their engineers, that the foam debris strike was not a threat to the safety of the mission?
3. How could NASA have forgotten the lessons of Challenger?
4. What should NASA do to minimize the likelihood of such accidents in the future?

Although the CAIB's comprehensive report raised important questions and offered answers to some of them, it also left many major questions unanswered (Starbuck and Farjoun, 2005).

1. Why did NASA consistently ignore the recommendations of several review committees that called for changes in safety organization and practices?
2. Did managerial actions and reorganization efforts that took place after the Challenger disaster contribute, both directly and indirectly, to the Columbia disaster?
3. Why did NASA's leadership fail to secure more stable funding and to shield NASA's operations from external pressures?

By examining, with respect to the Columbia disaster, the case of NASA as an organization, one can try to extract generalizations that could be useful for other organizations, especially those engaged in high-risk activities—such as nuclear power plants, oil and gas, hospitals, airlines, armies, and pharmaceutical companies—and such generic principles may also be salutary for any kind of organization.

The CAIB (2003) report recommended developing a plan to inspect the condition of all RCC systems, the investigation having found the existing inspection techniques to be inadequate. RCC panels are installed on parts of the shuttle, including the wing leading edges and nose cap, to protect against the excessive temperatures of reentry. They also recommended that taking images of each shuttle while in orbit should be standard procedure as well as upgrading the imaging system to provide three angles of view of the shuttle, from liftoff to at least SRB separation. "*The existing camera sites suffer from a variety of readiness, obsolescence, and urban encroachment problems.*" The board offered this suggestion because NASA had had no images of the Columbia shuttle clear enough to determine the extent of the damage to the wing. They also recommended conducting inspections of the TPS, including tiles and RCC panels, and developing action plans for repairing the system. The report included 29 recommendations, 15 of which the board specified must be completed before the shuttle returned to flight status, and also made 27 "observations" (CAIB, 2005).

11.5 **FTA AND RBD**

The fault tree analysis (FTA) diagram in Figure 11.2 and the reliability block diagram (RBD) in Figure 11.3 show that in order to avert such a disaster, it would have been important to have improved the reliability of the system (as evidenced by the damage of the TPS), support from management, and safety culture. All three factors contributed simultaneously to the disaster and hence have been linked using an AND gate in the FTA and, accordingly, a parallel structure in the RBD.

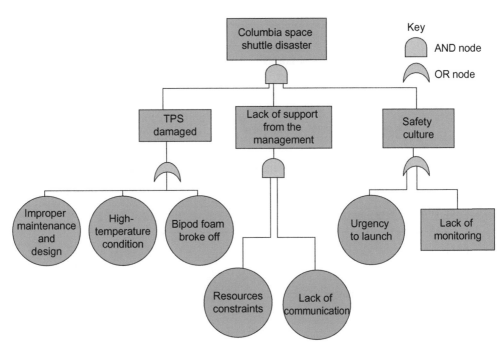

■ **Figure 11.2** FTA of the Columbia disaster.

11.6 **GENERIC LESSONS**

When reflecting on the Columbia disaster and the three dimensions of learning from failures (feedback to design, using analytic techniques, and extracting generic lessons) as outlined in Chapter 1, it is clear that there is evidence that the three dimensions have been addressed here as given in Table 11.1. The idea of generic lessons is that they focus on observations that are not just related to NASA but also to many other organizations. Generic lessons also have the feature of borrowing lessons that are known in one field and exporting them to others. For example, there may be lessons known in the maintenance engineering discipline which will be of benefit in safety engineering.

It is interesting to note, however, the striking similarities between the Challenger and the Columbia disasters. According to the CAIB (2003) report, on the Columbia accident *"Both accidents [Challenger and Columbia] were 'failures of foresight' in which history played a prominent role."* Also, the board found many *"echoes of Challenger"* in the decisions that led to the Columbia tragedy. When reading the reports on

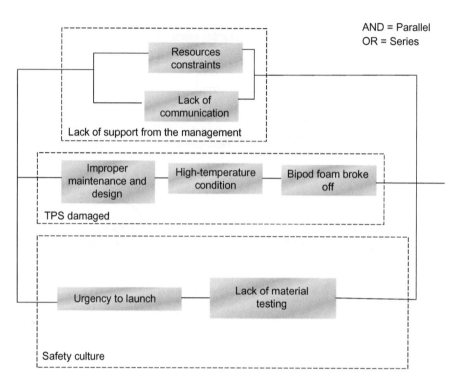

■ **Figure 11.3** RBD for the Columbia disaster.

Table 11.1 The Three Dimensions of the "Learning" Framework as Illustrated by the Analysis of the Columbia Disaster

Feedback from users (maintenance) to design.	Safety aspects of on flight monitoring and repair.
Incorporation of advanced tools in innovative applications, such as FTA and RBD.	a. From FTA: It is important to improve the reliability of the system, support from management, and safety culture.
	b. From RBD: Importance of simulation to avert physical damage of the TPS system (represented by a relatively high number of boxes in series structure).
Generic lessons and fostering of interdisciplinary approaches.	Decentralization and independence of safety budget.
	Recertification prior to operation.
	Provision of adequate resources for long-term programs to upgrade design and reliability of systems.

Source: Adapted from Labib and Read (2013).

both Challenger and Columbia, it is observed that *"a flawed decision-making process"* has been cited as the primary causal agent as well as *"communication failures,"* in both cases.

According to Buljan and Shapira (2005), one of the sobering aspects of analyses of the Columbia accident was that the decision to launch Columbia without ascertaining the proper functioning of the heat insulation was in some ways a replication of the disastrous decision to launch the Challenger. They suggested that it would be assumed that, having gone through the Challenger disaster, NASA would have assimilated the many recommendations of the subsequent investigation, and that its decision-making processes would be influenced by what it had learned. They then hypothesized that similarities between the Challenger and the Columbia disasters suggested that there were obstacles to organizational learning, some of which may have been embedded in the incentives and penalties that organizations employ to motivate their managers. They then concluded that both accidents looked very similar and were caused by human error, and not by force of nature, and suggested that learning did not occur.

Mahler and Casamayou (2009) suggested that, given the similarities in the causes of the Challenger and Columbia disasters, there may be evidence that NASA was "not a learning organization." They explained that, *"In essence, organizational learning is about the astuteness of the organization, and the honesty and curiosity of its members in uncovering problems."*

This is the crux of what makes the study of learning at NASA so interesting: It is an agency founded on curiosity and honesty—curiosity about our world and the universe pursued through rigorous scientific methods and disciplined engineering practices. Yet many observers outside NASA (investigators, overseers, academics, journalists), and indeed many inside the agency, have concluded that it has been unable or unwilling to apply this inquisitiveness and scrupulousness to itself as an organization. The unfortunate precedent of the Challenger disaster in 1986 provides a rich opportunity to compare two well-documented accidents and consider how NASA developed over time.

According to Vaughan (1996, 2005), both disasters were due to technical issues (O-ring blow-by in the case of Challenger, and foam shedding in the case of Columbia) where each has been experienced many times previously, but those occasions have been "normalized," despite individuals in the organization having expressed concerns.

Vaughan (2005) stated that there are three factors, combined in system effects, that produced both *Challenger* and *Columbia*: a decision-making pattern normalizing technical anomalies, creating a cultural belief that it was safe to fly; a culture of production that encouraged continuing to launch rather than delay while a thorough hazard analysis was conducted; and structural secrecy, which prevented intervention to halt NASA's incremental descent into poor judgment. She then claimed that the amazing similarity between the organizational causes of these accidents, 17 years apart, raises two questions: Why do negative patterns persist? Why do organizations fail to learn from mistakes and accidents?

There were organizational problems highlighted by both commissions that were charged with the investigation of the disasters (the Rogers Commission (1986) for Challenger and the CAIB (2003) for the Columbia), and yet, according to Leveson et al. (2005), the same organizational problems that created the Challenger accident were implicated in the Columbia accident.

The principal generic lesson that can be learnt from Columbia is perhaps about the concept of "unlearning." According to Mahler and Casamayou (2009), there are three categories of approaches to the lessons that organizations can derive from disasters: (1) not learned, (2) learned only superficially, or (3) learned and then subsequently unlearned. The idea of learning and then unlearning is of particular interest, as the concept of unlearning has received little attention in the literature. An example of "unlearning" at NASA concerns the organizational structure where, according to the Rogers report (1986), one of the reasons for the Challenger disaster was the existence of rivalry among centers. Different organizational cultures, lack of responsibility and accountability, and cross-center communication issues created information and coordination problems that led directly to the accident. After the shuttle program was restructured in a centralized configuration, which continued until the 1990s, but then reorganization occurred and the system reverted to one of different centers, which was intended to improve efficiencies; a policy of "*faster, better, and cheaper*" according to Woods (2006). This policy of coping with pressures created conditions favoring the Columbia disaster.

Leveson et al. (2005) suggested that there are some basic generic principles important to providing an effective system safety organizational structure:

1. System safety needs a direct and influential link to decision makers (influence and prestige).
2. System safety needs independence from project management (but not engineering).

3. System safety needs direct communication channels to most parts of the organization (oversight and communication).

In the aftermath of Columbia, there was evidence of that the preceding objectives were not achieved. For example, regarding influence and prestige, the CAIB (2003) report concluded that system safety engineers were often stigmatized, ignored, and sometimes even actively ostracized. It stated that *"Safety and mission assurance personnel have been eliminated [and] careers in safety have lost organizational prestige."* Whereas regarding independence, it noted that *"Given that the entire safety and mission assurance organization depends on the Shuttle Program for resources and simultaneously lacks the independent ability to conduct detailed analyses, cost and schedule pressures can easily and unintentionally influence safety deliberations."* Finally, as regards oversight and communication, it is found that *"organizational barriers … prevented effective communication of critical safety information and stifled professional differences of opinion."* It was *"difficult for minority and dissenting opinions to percolate up through the agency's hierarchy."*

Yet further evidence of unlearning at NASA occurred when coping with budgetary pressures led to staff cutbacks, which ultimately led to the disappearance of organizational memory. Kletz (2009) rightly claims that people have memories whereas organizations don't. It is therefore important to develop innovative mental models that can contribute toward knowledge retention within the organization. It is hoped that the use of modeling in the form of FTA and RBD can contribute toward achieving this goal.

Another interesting lens through which to examine NASA's situation was offered by Heinmann (2005), who proposed that organizations are often in the position to commit two types of error: (1) implementation of the wrong policy (an error of commission) and (2) failure to act when action is warranted (an error of omission) as given in Table 11.2.

The same dilemma occurred in the decision to launch the shuttle. The launch of a mission that was actually unsafe would be a Type I error. It was such an error that caused NASA to lose a space shuttle, worth more than $2 billion, as well as the death of the seven astronauts, a loss which cannot be quantified.

On the other hand, if NASA had decided not to launch a mission that was technologically appealing, it would have committed a Type II error. A Type II error would have meant a waste of resources in setting up the launch (e.g., of the propellants in the external fuel tank and of liquid

Table 11.2 Two Types of Possible Error in NASA

		Proper Course of Action	
		Launch	*Abort*
NASA's decision	*Launch*	Correction decision Mission successful	Type I Error: Accident occurs; Possible loss of life and/or equipment
	Abort	Type II Error: Missed opportunity Wasted resources	Correction decision Accident avoided

Source: *Adapted from Heinmann (2005).*

hydrogen and liquid oxygen, in total valued at approximately $500,000). Moreover, NASA would have missed a rare opportunity to carry out scientific research.

It is apparent that, in both the Columbia and the previous Challenger accidents, the Type I failure that caused them was far more devastating than a Type II failure would have been. There are other cases however, where a Type II error could be the more expensive: the deciding whether, for example, an important medical drug, that could save many lives, should be tested.

Another interesting lens of analysis was offered by Ocasio (2005) who examined the interplay between language and culture in the Columbia disaster. He observed that within the culture of the space shuttle organization, the meaning of "safety of flight" was ambiguous and people viewed safety as a minimal constraint to satisfy rather than a goal to raise.

A further lens of analysis has been proposed by the author and a colleague (Labib and Read, 2013) who introduced 10 generic lessons of learning from failures. In the next section an attempt is made to adapt those lessons, and map them, to the case of NASA in order to outline the unlearning process that occurred there, i.e., the increase of repetitive disasters within the same organization. As also proposed, the generic lessons will be grouped into their underpinning attributes of (i) priorities, (ii) responsibility and skills, and (iii) communication.

11.6.1 **Generic Lessons That Are Mainly Related to Setting Priorities**

11.6.1.1 *Too Much Belief in Previous Successes*

It is argued that experience with success can be counterproductive. In the Challenger case, Vaughan (2005) observed that according to the Commission, NASA leaders' belief in operational capability was reinforced by the space shuttle history, prior to Challenger, of 24 launches without a failure and to NASA's legendary "can-do" attitude, in which the agency always rose to the challenge, draining resources away from safety-essential functions to do it (Rogers Presidential Commission Report, 1986: 171–7). In their analysis of the Challenger disaster, Starbuck and Milliken (1988) observed that repeated successes nurture *"complacency, confidence, inattention, routinization, and habituation."* They also suggested that *"successes may induce engineers and managers to attempt to fine-tune a sociotechnical system—to render it less redundant, more efficient, more profitable, cheaper, or more versatile. Success breeds confidence and fantasy. When an organization succeeds, its managers usually attribute this success to themselves, or at least to their organization, rather than to luck."* Richard Feynman interpreted NASA's behavior as that *"after each successful flight, NASA's managers thought 'We can lower our standards a bit because we got away with it last time' "* (Rogers Presidential Commission, 1986: I-148).

11.6.1.2 *Coping with Growth*

The ability, or inability, to cope with a high rate of growth is an important factor that can contribute to disastrous failures. Vaughan (2005) explained that NASA's institutional environment and the culture of production led to the shuttle being perceived as an operational vehicle, not an experimental system, which led to more emphasis on operation and growth and less on safety.

The report of the Columbia disaster argued that the shuttle program concerns about the *Columbia* foam strike were not about the threat it might pose to STS-107, but about the threat it might pose to the schedule (CAIB, 2003). Analyzing the disaster, Buljan and Shapira (2005) concluded that while engineers focus on safety, managers focus on reaching the organization's target.

11.6.1.3 *Misconception of Fashionable Paradigms*

The misconception of fashionable paradigms can be a reason why man-made disasters happen. According to Vaughan (2005) and Woods (2006), the initial surge in post-*Challenger* funding was followed by budget cuts in the 1990s, causing the new NASA Administrator, Daniel Goldwin, to introduce new efficiencies and smaller programs with the slogan, "faster, better, cheaper," a statement later proved to have strong cultural effects and an impact on the Columbia disaster. The CAIB (2003) report stated: *"When paired with the 'faster, better, cheaper' NASA motto of the 1990s and cuts that dramatically decreased safety personnel, efficiency becomes a strong signal and safety a weak one."*

11.6.2 Generic Lessons That Are Mainly Related to Responsibility and Skills

11.6.2.1 *The "I Operate, You Fix" Attitude*

Fragmentation and demarcation of skills and responsibilities can be a reason for disasters. Vaughan (2005) noted that unless NASA engineers defined something as a serious problem, it was not brought to the attention of safety personnel. NASA had increased reliance upon contractors for safety, relegating many NASA technical experts to desk-job oversight of contractor activities. At the same time that this strategy increased NASA dependence on contractors, it undermined in-house technical expertise.

In the aftermath of the Challenger disaster, Starbuck and Milliken (1988) concluded that *"the traditional differences in the responsibilities of engineers and managers give their interactions an undertone of conflict and make learning partly a process of fine-tuning the probabilities of success. Fine-tuning gradually makes success less and less likely."* In a televised documentary about the Challenger disaster, Jerlad E. Mason, who was senior manager at Thiokol—the contractor responsible for the SRB for NASA at the time of the Challenger mission—urged one of his engineering managers to *"take off your engineering hat and put on your management hat"* in order to convince him to vote for a shuttle flight recommendation which was against his engineers initial recommendation of no-fly, which ultimately sealed the fate of Challenger.

11.6.2.2 *Skill Levels Dilemma*

In the maintenance function, the designer of the machine is not usually the one who fixes it when it fails and, surprisingly, might not even have the ability to do so Vaughan (2005) observed that safety at NASA *"suffered from personnel cuts and deskilling as more oversight responsibility was shifted to contractors in an economy move, and it [NASA] had no ability to independently run tests that might challenge existing assessments.*" Also, the Commission found that in-house safety programs were dependent upon the parent organization for funding, personnel, and authority (CAIB, 2003). This dependence showed when NASA reduced the safety workforce, even as the flight rate increased.

11.6.3 **Generic Lessons That Are Mainly Related to Communication**

11.6.3.1 *No News Is Good News*

In old-fashioned maintenance, the prevailing attitude is to make minimal effort for prevention, and the main mode of work is corrective, as opposed to proactive. One famous saying is *"If it ain't broke, don't fix it."* This implies a passive attitude toward performing any preventive activity, a wait-and-see situation. Vaughan (2005) noted that there was a culture of structural secrecy at NASA, by which she meant the way in which organizational structure and information dependence obscured problem seriousness from people responsible for oversight.

11.6.3.2 *Bad News Bad Person*

In old-fashioned organizations, the manager always prefers to hear good news. Someone who brings bad news about a malfunction, or even the expectation of it, is considered an under-performer. Vaughan (2005) posed an interesting question: why had no one responsible for safety oversight acted to halt NASA's two transitions into disaster? She then claimed that individual secrecy—the classic explanation of individuals trying to keep bad news from top management was the reason.

11.7 **CONCLUDING REMARKS**

As mentioned previously, it was observed by Starbuck and Milliken (1988) that repeated success nurtures confidence, which may in turn lead to accusations of complacency and inattention. These could be felt to be derogatory, prompting punishment of any colleague who might voice concerns.

A final reflection on this disaster: it is crucial to keep seeking the right questions, and to obtain the right data, prior to taking any important decisions. Vaughan (2005) made an interesting observation: *"it showed how both the history of decision making by political elites and the history of decision making by NASA engineers and managers had twice combined to produce a gradual slide into disaster."* This shows the importance of mental models as offering a language that can help to both retain knowledge and facilitate better communication for decision making.

11.8 **CRITICAL COMMENTARY SECTION**

It is noted that not everybody agrees with the findings and conclusions of each chapter. So at the end of some of the chapters there is an alternative view, which is not intended to undermine the issues discussed but rather present other perspectives. The ideas included do not necessarily represent the author's own view, but they are worth consideration and worth initiating some reflection and debate.

CRITICAL COMMENTARY: SAFETY IS CONSIDERED BY SOME MANAGERS AS A VERY EXPENSIVE DRAIN ON THEIR RESOURCES

True though this may be, safety being only too often considered to be a non-issue (i.e. regarding events that have not happened), the whole picture needs to be examined and consideration given to the 'what if' scenarios. Again, a good way of expressing this was provided by Trevor Kletz in the 2008 CSB safety video "Anatomy of a Disaster"

There's an old saying that if you think safety is expensive, try an accident.
Accidents cost a lot of money. And, not only in damage to plant and in claims for injury, but also in the loss of the company's reputation.

It is difficult to quantify the loss of reputation but it will probably dwarf any of the costs of ensuring safety.

Titanic, the Unsinkable Ship That Sank

12.1 CRUISING

Cruising saw its beginning in 1801 when the first steam-driven vessel, the Charlotte Dundas, was launched in Scotland (Cartwright and Baird, 1999). According to Attard (2013), in the 1900s, cruising started taking a different shape. Iron hulls and turbine-driven propellers fitted with large reciprocating steam engines were developed, and ships got bigger. In 1907, Cunard launched two ships, the Mauritania and Lusitania, both of which were over 30,000 Gross Registered Tonnage (GRT). White Star responded when they introduced Olympic, which was 46,320 GRT (Cartwright and Baird, 1999). Producing large luxurious ships seemed to be the new thing to do, and that is when the Titanic (46,328 GRT) was launched.

Cruising was an expensive holiday and was only available to the selected few. Safety was a dominant factor in the cruising business, and everyone believed that these ships were indestructible. This was not the case. On its maiden voyage from Southampton, the Titanic collided with an iceberg and sank in the Atlantic Ocean with a loss of over 1500 (passengers and crew) (Cartwright and Baird, 1999).

12.2 WHAT HAPPENED

The sinking of the Titanic on April 15, 1912, has come to be one of the most famous disasters ever. At the time it was regarded as the world's biggest and most luxurious ship. About 900 ft long and 25 stories high, it weighed 46,000 tons (Division, 1997). It hit an iceberg at such high a speed that the resulting tear in the hull, below the waterline, ruptured 5 of its 16 watertight compartments, and it sank 2 h and 40 min after the collision. Passengers and crew on board totaled 2228 (Figure 12.1).

Learning from Failures. DOI: http://dx.doi.org/10.1016/B978-0-12-416727-8.00012-6

■ **Figure 12.1** The Titanic.

12.3 **LOGIC OF THE TECHNICAL CAUSE OF THE DISASTER**

It is believed that many factors led to the disaster, but most notably they were attributed to wrong decisions, the *low visibility* of a moonless night, limited resources insufficient for detection of icebergs (a lack of technology and equipment), metal failure (technology not available for *quality control and testing*), *inefficient evacuation procedure* and *insufficient lifeboats* (they could carry only 1178 people), and too great a belief in the ship's unsinkability.

12.4 **CONSEQUENCES AND SEVERITY**

A loss of 1517 people (Figure 12.2), loss of a ship valued at $7.5 million (in 1912 monetary terms, which roughly equates to $200 million in 2012), severe damage to the reputation of the White Star Line and a massive increase in insurance premiums.

12.5 **FTA AND RBD FOR THE DISASTER**

The fault tree analysis (FTA) shown in Figure 12.3 shows that there were three main causative factors, namely (a) insufficient means of passenger evacuation, including too few lifeboats, not enough time to evacuate, and lack of emergency procedures; (b) iceberg collision due to poor visibility or high speed; and (c) hull failure due to either inappropriate material or manufacturing processes. Note, as mentioned in the previous chapters,

■ Figure 12.2 Consequence of the Titanic disaster.

that the OR logic gate denotes whether one factor was sufficient, and the AND logic gates signifies that all factors are needed to happen simultaneously.

The initial analysis was carried out by a group of students at the University of Southampton—the Titanic's port of departure. (The students selected some interesting words in their initial modeling, such as "No binoculars," rather than "No early warning devices" and "Bad outlook" rather than "Bad weather conditions." I liked their choice of words, as they were simple and appealing and a bit funny, and so I have left them in.)

The reliability block diagram (RBD) of Figure 12.4 was derived from the FTA, each AND gate in the FTA being represented by a parallel structure in the RBD and each OR gate by a series structure, and we normally start modeling from the top of the tree and then cascade downward. In general, an AND gate is a more reliable configuration than OR gate. From Figure 12.3, it can be seen that the second parallel line "Insufficient

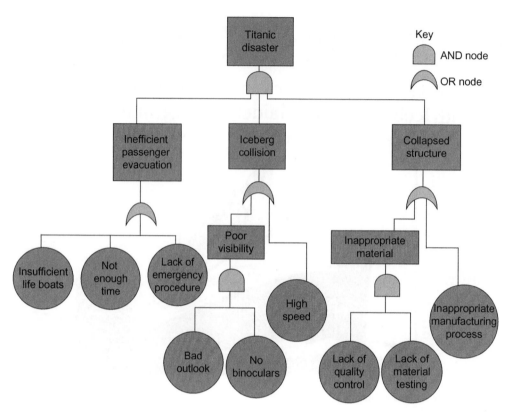

■ **Figure 12.3** FTA of the Titanic disaster.

passenger evacuation" indicates the most vulnerable area, all the boxes being arranged in series, whereas in the other lines there are some parallel structures. Here we are assuming that boxes carry equal weights. (The actual vulnerability of a line will of course depend on the actual probability values of the blocks included and not just on the way they are arranged. Although such probability values would of course enhance the calculation, it is often difficult to acquire or assess them.)

12.6 **PROPOSED IMPROVEMENTS AND GENERIC LESSONS**

After carrying out a disaster analysis using FTA and RBD, there is a need to reflect upon what could have been done differently to prevent the particular incident. A significant factor in the Titanic disaster was the lack of human adherence to safety procedures, which included

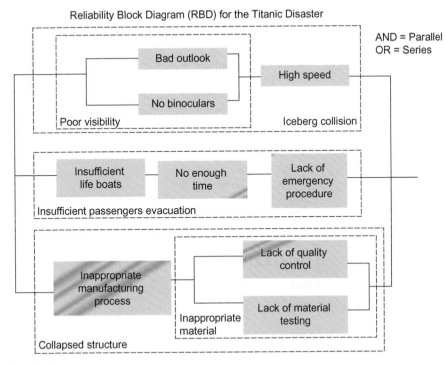

■ Figure 12.4 RBD of the Titanic disaster.

nonconcurrence of decisions and inadequate training regarding safety procedures (especially with respect to passenger evacuation). Another important factor was the availability of technology and equipment. This could have included real-time observation and safe sailing speed. Regarding poor design, the accident showed that there was a need for redesigning ships to remain afloat after major impact and to review the best available technologies to achieve this (Table 12.1).

Recent research has been carried out by Attard (2013) about the percep-tions of people regarding cruise liner disasters (like the, most recent, Costa Concordia accident). She concluded that such disasters had little effect on people's perception about cruising, especially since these events are far from being a daily occurrence. Health and safety issues have been triggered however, and through amended regulations cruising has become safer than ever. Nevertheless, the trend for ever larger ships, increasing the possibility that should a disaster happen it will have larger consequences for the industry and in terms of the passengers involved.

Table 12.1 The Three Dimensions of the "Learning" Framework as an Outcome of the Titanic Disaster Analysis

Feedback from users (maintenance) to design.	Number of lifeboats and evacuation and emergency procedures.
Incorporation of advanced tools such as FTA and RBD	a. From FTA: (a) Insufficient means of passenger evacuation, including too few lifeboats, not enough time to evacuate, and lack of emergency procedures; (b) iceberg collision due to poor visibility or high speed; and (c) hull failure due to either inappropriate material or manufacturing processes. b. From RBD: insufficient passenger evacuation was the greatest vulnerability.
Generic lessons and fostering of interdisciplinary approaches.	Safety procedures. Use of technology. Redesign of hull and watertight compartments.

Table 12.2 Landmarks of Cruising History

Year	Event
1801	The first steam-driven vessel in Europe was launched: the Charlotte Dundas in Scotland.
1807	The first steam-driven vessel in America, the Cleremont by Foulton, founded in New York.
1838	This year marked the first Atlantic Ocean steam crossing.
1881	First pleasure cruise launched.
1907	Introduction of larger ships (over 30,000 GRT).
1912	The Titanic set off for its maiden voyage. Up until then, cruising was considered a safe thing to do, but this was marred when the Titanic collided with an iceberg and sank.
1939	The start of the Second World War brought about the cessation of pleasure cruising.
1960	The progression in air transport saw cruise companies struggling.
1990	Mega ships started to be built. The first was the Sovereign of the Seas, which weighed 73,000 GRT.
2000 to date	Ships have kept on increasing in size, and in facilities on board. The largest ship to date is the Allure of the Seas (225,282 GRT).

Table 12.2 gives the landmarks of cruising history and the impact of the Titanic disaster. It is adapted from the work of Attard (2013), who cited sources from Cartwright and Baird (1990) and Xuereb (2008).

Table 12.3 has been adapted by the author from a compilation by Attard (2013) of cruise liner disasters which have had a major impact on the industry.

Table 12.3 Major Disasters in the Cruising Industry

Name of Disaster	Description
RMS Titanic	This is probably the best known disaster when it comes to cruise liner incidents. The Titanic set out on its maiden voyage on the April 10, 1912, ending it 5 days later when it struck an iceberg on its way to New York. The Titanic was known as the unsinkable ship. However, only 3 h after it hit the iceberg, it was lying at the bottom of the Atlantic. Out of 2228 passengers and crew, 1517 lost their lives, out of which 685 were crew and 53 were children.
S.S. Eastland	This ship was known as the "Speed Queen of the Great Lakes" and was meant to take the Hawthorne Works employees and their families and friends across to Michigan. Unfortunately the S.S. Eastland never left Chicago but instead it listed on its side on the July 21, 1915, while still docked. There were over 2500 passengers on board. In all 844 people were killed, out of which 2 were crew members.
The Herald of Free Enterprise	The Herald was a "Spirit Class" Ferry, 132 m long, capable of carrying up to 1400 passengers. It sank on March 6, 1987 just outside the port of Zeebrugge on a voyage to Dover. It capsized and sank within 2 min of water entering the car deck. Of the 459 people on board, 188 were killed. The Coroner's inquest found a verdict of unlawful killing because of gross negligence. P&O European Ferries faced a charge of corporate manslaughter, and seven individual employees faced charges of manslaughter, though none were prosecuted.
Royal Pacific	Departed Singapore on the evening of August 21, 1992, with 534 people on board, collided with Terfu 51, a Taiwanese fish trawler, due to lack of visibility. Three people drowned and six were reported missing.
Seabourne Spirit	In November 2005, after sailing for more than a day with no other ship in sight, the Seabourne Spirit was fired upon by armed people on boats off the coast of Somalia. Passengers moved to the central lounge for safety as grenades exploded. When pirates tried to board the ship, the captain manoeuvred it further out to the sea, making it impossible for the pirates to board. No one was injured. However, in that year more, than 32 pirate hijackings occurred in the area and they are nowadays not an uncommon occurrence, especially in the Red Sea, Gulf of Aden, and Indian Ocean.
Norwegian Dawn	On the April, 10, 2005, the Norwegian Dawn was on its way to New York after a cruise in the Bahamas. It encountered heavy weather and seas about 120 miles away from the coast of South Carolina and was then hit by a 70 ft tall wave which made the ship go from one side to another. Cabins on Decks 8 and 9 sustained the major damage, more than 60 being flooded and 4 passengers required treatment for cuts and bruises.
Star Princess	In March 2006, at around 3 am, fire broke out in a passenger's cabin, caused by a cigarette which was thrown from upper balconies. The captain sounded the general emergency alarm, waking up all the passengers who were grouped at their muster stations after making their way through smoky corridors. 150 cabins were completely damaged, another 100 sustained smoke damage, 13 people suffered from smoke inhalation, one of whom died.
Celebrity Mercury	During a Caribbean cruise in 2010, 350 passengers out of 1800 fell ill with upset stomach, vomiting, and diarrhea, which could have been due to an outbreak of Norovirus, that spreads in closed environments. Due to the large number of people who fell ill, more medical staff boarded the ship at the next port of call.
Costa Concordia	This is probably the best known cruise liner disaster after that of the Titanic. On January 13, 2012, the ship left Civitavecchia near Rome, on the way to its first port of call. Carrying more than 4250 (passengers and crew), the ship struck a rock off the Italian Island of Giglio, ripping a 50 m gash on the left-hand side of its hull. This led to immediate flooding and electrical power cuts which impeded the evacuation process. According to international

Continued

Table 12.3 Major Disasters in the Cruising Industry *Continued*

Name of Disaster	Description
	marine laws, evacuation should not take more than 30 min. However, evacuation took over 6 h. Thirty lives were lost, including that of a 6 year old, and two people who are still missing. This incident reminded everyone of the Titanic and induced a some uncertainty in the cruise liner industry.
Thomson Majesty	In February 2013, five crewmen were killed and three were injured, while checking a lifeboat. The accident happened when one of the lifeboats hoist cables snapped during a drill, which sent the boat plummeting down 65 ft into the sea from the top deck. It is said that the lifeboat overturned when it hit the water leaving the crew members trapped underneath.

The accident that is most similar to the Titanic disaster is that of the Costa Concordia. Both collided (one with an iceberg, one with rock), and both incurred loss of life, much more though in the case of the Titanic as it was in the middle of the ocean, with freezing waters. On the other hand, the Costa Concordia was very close to shore and a number of passengers swum to land and rescue arrived quicker (Attard, 2013). According to Attard, disasters like that of Costa Concordia do not happen often and are almost unheard of. The Costa Concordia incident could be listed as the second biggest disaster after the Titanic and is probably the biggest such disaster that so far the present generation remembers. These incidents however would have a negative aftermath on the cruise industry, and even companies not directly concerned would be affected.

One last word: with the increase of size in all means of transportation, most notably in aeroplanes and ships, the consequence of any disaster on them becomes more significant. So perhaps size limitations might be one way of lowering the long-term consequences of disasters.

Part 3

Generic Lessons, Other Models of Learning from Failures and Research Directions

Introduction to the Concept of the Generic Lesson as an Outcome of Learning from Failures

13.1 WHY FAILURE CAN BE THE BEST THING TO HAPPEN

This may sound rather cynical but, unfortunately, there is an element of truth in it: "*Some of the world's greatest disasters could have been avoided if those behind them had experienced more failure, according to research published this week.*" This is an extract from a press release following the publication of a paper by the author and a colleague, on learning from failures (Labib and Read, 2013) in the journal of *Safety Science*. Shortly afterward, the author became a minor celebrity, briefly, when the paper was featured in about 10 newspapers and he was asked to give a radio talk about it on the BBC. The paper argued that there is evidence that some of the lessons gained from major disasters have not been learned by the very same organizations in which the disasters had occurred. It was suggested that the sinking of the Titanic, the loss of the space shuttles Columbia and Challenger, two BP oil refinery explosions with huge loss of life, and Toyota's international recall of more than 8 million of its cars all had in common an inflated degree of confidence. Avoiding overconfidence is among a list of 10 "generic lessons," based on studies of high profile disasters designed to help organizations and managers to understand the reasons for such disasters.

In this book it has been argued that organizations learn more effectively from failures than from successes, but that they vary in learning from them. Also, they often learn vicariously from the failures and near-failures of other organizations. A lack of failure, on the other hand, can lead to overconfidence and "blindness" to the possibility of problems.

Learning from Failures. DOI: http://dx.doi.org/10.1016/B978-0-12-416727-8.00013-8

Some managers and organizations see their role as akin to "rearranging the deckchairs on the Titanic," but disasters, when you study them, are often built on similarly futile exercises. It has also been argued that for organizations to successfully avoid disasters they need to undertake risk and reliability analyses. Policy makers also need to balance their punishment and incentive systems and ensure that they are proportional to the significance of hazardous incidents.

Failures in general, and disasters in particular, can engender a blame culture that can act as a barrier to learning from failures, but it is important to note that human beings are naturally programmed to learn, whereas organizations are not.

Throughout this book 10 generic lessons, for avoiding major failures, have been identified, based on the root causes of major problems and an understanding of how those problems evolved over time. In the following section, a brief account will be given about these lessons, drawing on some similarities with other disciplines such as maintenance management and reliability engineering.

13.1.1 Lesson 1: Too Much Belief in Previous Successes

Experience with success can be, and has been, counterproductive. Too much belief in the "unsinkability" of the Titanic meant it set sail with too few lifeboats and those that were installed were there only to rescue people from other ships. It sank, in 1912, with the loss of 1300 lives despite being perceived, at the time, *"as the safest ship ever built"* (The Sinking of the Titanic, 2000). It was this false perception that led to the fatal error of providing insufficient lifeboats.

NASA's confidence in its space shuttle program led them to ignore warning signals about the O-ring damage, due to cold weather, prior to the Challenger disaster in 1996, and again about the fuel tank foam losses prior to the Columbia disaster in 2005. According to the investigation report, NASA's safety culture had become reactive, complacent, and dominated by unjustified optimism. In this context, the culture was criticized with regard to inadequate communication and setting priorities.

13.1.2 Lesson 2: Coping with Growth

The inability to cope with a high rate of growth may be another factor that can contribute to disastrous failures. Toyota has been regarded as the company against which other organizations benchmark their standards, as evidenced in the best-selling management book, *The Machine That Changed the World* (Womack et al., 1990), which is based on the Toyota Production System (TPS). In 2010, however, failures led to the massive recall of Toyota vehicles in the United States, Europe, and China. It has been claimed that this was attributed to the company's inability to cope with its rate of growth, which was such that it overtook GM in April 2007 to become the world leader in car market share.

Could BP have encountered same circumstances of rapid growth, given the soaring of its share prices, prior to the Deepwater Horizon accident in 2010? One would expect that the magnitude of catastrophic failures would be inversely proportional to the resources available. So, in good times, resources would be more available and hence more investment would go into safety measures. However, it appears that the relationship between a company's prosperity and its safety is probably of an inverted U-shape, where the two extremes of bad or good times are not good news as far as safety is concerned. It is difficult to assess this hypothesis but history appears to teach us that in bad times, resilience, and even faith, tends to be stronger than in good times. In this context what matters is how we set priorities regarding the allocations of resources for safety measures.

13.1.3 Lesson 3: Misunderstanding Fashionable Paradigms

The misconception of fashionable paradigms might be one of the reasons why man-made disasters happen. Such misunderstanding can come in many forms, including, for example, misunderstanding of "lean" management. The concept of lean thinking, or lean manufacturing, is often misinterpreted as a cost cutting exercise, "taking the fat out of the meat." Organizations, in an attempt to identify and minimize waste, can go too far and end up unknowingly sacrificing safety. Cutting cost is just one aspect of lean thinking; it is also about adding value and streamlining a process. We have seen this before when business process engineering (BPR) was a fashion, and it ended up being mocked as "bastards planning redundancies." Could any of the organizations discussed in this book have sacrificed safety in the name of "lean"? Misconception of paradigms has always been a dangerous affair (Alvi and Labib, 2001). For a good analysis of what is a fad and what makes a paradigm, see Towill (1999). The issue here is about how such paradigms are communicated within an organization in a way that does not lead implicitly to sacrificing safety measures.

13.1.4 **Lesson 4: Legislation**

Discussing the Bhopal disaster in 1984, where more than 3000 died, Chouhan (2005) made a comparison between the two sister plants owned by the same company, Union Carbide, and argue that the significant differences between the West Virginia plant and the Bhopal plant showed the callous disregard of the corporation for the people of developing countries. Twenty-six years after the disaster, the officials responsible were prosecuted, but there seems to be a need for something similar to the three strikes and outlaw, at a corporate level in terms of serious breaches of safety. The United Kingdom's Corporate Manslaughter and Corporate Homicide Act of 2007 is a good step in this direction; for the first time, organizations can be found guilty of corporate manslaughter as a result of serious management failures resulting in a gross breach of a duty of care. Also in the United Kingdom, "whistle blowing" legislation has been introduced (The Public Interest Disclosure Act of 1998) which is designed to protect workers who disclose information about dangers to health and safety in their organization (Flin, 2006). In this context, legislation can guide and control responsibilities for, and priorities of, safety.

13.1.5 **Lesson 5: The "I Operate, You Fix" Attitude**

In old-fashioned "traditional" maintenance, a prevailing concept among operators is "*I operate, You fix.*" In other words, maintenance is totally the responsibility of the maintenance department and operators should deal only with the operation of their own machines. When dealing with a disastrous situation, this attitude frequently means that most people feel the responsibility for dealing with a disaster lies with someone else. But it is important that everybody, especially leaders and senior management, are aware that a disaster is not just "another issue" and that their direct involvement is necessary.

Comparing the maintenance function with the safety function is an interesting analogy. For example, the old-fashioned attitudes in maintenance (which used to be the prevailing ones in the West), and that are shown in Figure 13.1, can be compared to the new attitudes promoted by the Japanese total productive maintenance (TPM) philosophy. In old-fashioned maintenance, skills are fragmented (depicted in different colors in the figure), so the operator would claim "*I operate, you* [meaning maintenance people] *fix it.*" Then the maintenance engineer would claim, in return, "*I fix it, but you* [meaning designers] *design it.*" Finally, the designer would respond "*I design it, but you* [meaning operators] *operate it.*" This culture shifted responsibility, and the lack of ownership led to

- OLD Attitude:
  - " I operate it, you fix it "
  - " I fix it, you design it "
  - " I design it, you operate it "

- NEW Attitude:
  "We are all responsible
  for our equipment "

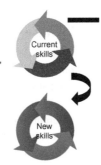

■ **Figure 13.1** Old versus new attitude in maintenance (but can also apply to safety). (For interpretation of the references to color in this figure legend, the reader is referred to the web version of this book.)

shifting the blame to someone else, or to another function within the business. On the other hand, the new attitude could be pictured as all in a single-colored circle, implying a culture of *"we are all in one ship, so we are all responsible, and if it sinks we all drown."* No attitude of "them" and "us," and ownership shared among all the team members. It is all about responsibilities.

13.1.6 **Lesson 6: No News Is Good News**

In old-fashioned maintenance, the prevailing attitude is to make minimal effort for prevention, and the main mode of work is corrective maintenance. One famous saying is "If it ain't broken, don't fix it." This implies a passive attitude toward performing any prevention activity, a wait-and-see situation. In most organizations, a preventive maintenance schedule could be well prepared but usually not implemented. The reason is that breakdowns take priority over preventive actions. In the context of disasters, it seems that the same attitude prevails, especially in the private sector, as organizations are.

13.1.7 **Lesson 7: Bad News, Bad Person**

In many organizations, the manager prefers to hear good news and anyone who brings bad news about a malfunction, or even about the expectation of it, is at risk of being considered an underperformer, and in some cultures the term used is a "whistle blower," sometimes perceived as a trouble maker. In the context of an evolving disaster, for example, a CEO, when asking about problems, might be swiftly told that everything is under control, in order to please him. Here, the issue is about communication as well as responsibility.

13.1.8 Lesson 8: Everyone's Highest Priority Is Their Own Machine

In traditional maintenance, every operator's highest priority is their own machine, and the one who shouts loudest gets his job done first. This lack of a systematic and consistent approach to setting priorities can be an important factor in a disaster. Setting priorities should attract the highest attention among different approaches to protecting against any potential disasters. Questions such as who should set priorities, what criteria should be considered, and how to prioritize the allocation of resources need urgently to be addressed. In this context, setting priorities is everything. The decision-making grid (DMG), described in Chapter 14, is a method used to prioritize machines (assets) on the basis of a combination of the severity of their failure (measured by downtime) and the frequency of those failures, provides various maintenance strategies in response to the relative state of the machines (Labib, 2004, 2008; Fernandez et al., 2003).

13.1.9 Lesson 9: Solving a Crisis Is a Forgotten Experience

It is often the case that the solving of a problem does not get recorded or documented, but it is beneficial to both organizations and individuals to be able to easily access databases of mistakes or near-misses. For example, in old-fashioned maintenance, it is often the case that the solving of a breakdown problem does not get recorded or documented, it being regarded as a bad experience that is best forgotten.

Imagine that you ask several job applicants to write their CV. It is most likely that they will all write about their achievements and none will write about their bad experiences or about any failures they had in their career, socially or academically. Nobody would be proud to mention them. On the other hand, modern maintenance techniques stress that a crisis is an opportunity for investigation, and failures should be well documented for future analysis. Unfortunately, in many near-miss situations, organizations, and people, do not reveal their experiences with failing equipment or mistakes. One reason for that might be the fear of losing lawsuits or insurance claims. Good examples of this can be found in the healthcare system (Barach and Small, 2000). It would, however, be beneficial, to both organizations and individuals, to be able to easily access databases of mistakes or near-misses (Kim and Miner, 2007), a matter of communicating them and of assigning of responsibility for storing and accessing the lessons learnt. A crisis is not the worst of failures, but not to have tried to learn from it is the true failure. Failure can be regarded as success when we learn from it.

13.1.10 **Lesson 10: Skill Levels Dilemma**

In the maintenance function, the designer of a machine is not usually the one who repairs it, and, surprisingly, might not even have the ability to do so. The skills needed to restore particular equipment involve diagnostics, logical fault finding, disassembly, repair, and assembly. Depending on the level of complexity of the equipment, and of the function that the equipment carries out, the necessary skill level can be determined. According to a survey conducted by McDonald (2006) of aircraft maintenance technicians, in approximately one-third of the tasks they carried out, the technicians reported that they did not follow the procedure in the maintenance manual. They reported that there were better, quicker, and even safer ways of doing the task than following the manual to the letter. McDonald argued that manuals themselves are not an optimum guide to task performance as they have to fulfill other criteria, such as being comprehensive and up to date. The question is: How to bring operator requirements to the forefront of the design process? or how to feedback the knowledge, skills, and experience of the operator, who deals everyday with the machine, to the designer?

In a crisis, accessing appropriate types and levels of skill presents a major dilemma because disasters tend to present multidisciplinary problems spanning various fields such as information systems, maintenance, decision making, and crisis and risk management, and hence there is a need for a synchronized multidisciplinary team approach (Labib, 1999). In summary, this is about communicating failures back to design and taking responsibility for storing and accessing all relevant information.

The next section outlines the underpinning principles that impact upon these generic lessons and discusses whether there can be a model that summarizes the whole concept.

13.2 **ATTRIBUTES OF THE GENERIC LESSONS**

The generic lessons discussed can be characterized by three main issues: (a) responsibility, (b) communication, and (c) priority. You will have noted that at the end of each generic lesson, these three keywords were used to summarize the context of the lesson.

There are at least five lessons that are attributed to each issue—see Table 13.1—where each column has at least five ticks. However, each

Table 13.1 Generic Lessons Learnt from Failures

Generic Lessons	Responsibility	Communication	Priority
Generic lesson 1—Too much belief in previous successes		✓	✓
Generic lesson 2—Coping with growth			✓
Generic lesson 3—Misunderstanding fashionable paradigms		✓	
Generic lesson 4—Legislations	✓		✓
Generic lesson 5—The "I operate, you fix" attitude	✓		
Generic lesson 6—No news is good news			✓
Generic lesson 7—Bad news bad person	✓	✓	
Generic lesson 8—Everyone's higher priority is their own machine			✓
Generic lesson 9—Solving a crisis is a forgotten experience	✓	✓	
Generic lesson 10—Skill levels dilemma	✓	✓	

generic lesson affects one or more of the three categories (columns). In this categorization, we assume that priority is defined in the broader sense. That is, that it covers areas of criticality and resource allocation. Take, for example, one of those categories, say "*Responsibility.*" One can find that, in terms of responsibility: "*Legislations,*" "*The I operate, you fix attitude,*" "*Bad news, bad person,*" "*Solving a crisis is a forgotten experience,*" and "*the skill levels dilemma*" are all issues that have been addressed.

All those lessons related to responsibility share a common theme: trying to answer the main question about who is responsible for doing what. One could similarly view lessons learnt in terms of communication and prioritization.

Communication covers links between employees within different functions in the organization as well as within the same function. It is also about links across the supply chain, between companies and their suppliers, and between companies and their customers. With regard to a disastrous situation, if any of those links is weakened, then possibly one, or a combination of the five features related to communication (see Table 13.1), would occur.

Prioritization can be addressed from two points of view: (a) mode of work and (b) degree of criticality. One view concerns the priority of the mode of work: e.g., preventive versus corrective modes as shown in the generic lesson "*No news is good news,*" or allocating resources in response to changes in the market, as shown in the generic lesson

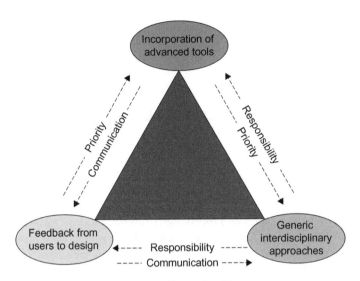

■ **Figure 13.2** A unified model of the process of learning from failures.

"*Coping with growth.*" Another view of priorities is that it involves a degree of criticality of equipment as shown in generic lessons "*Too much belief in previous successes*" and "*Everyone's higher priority is their own machine.*"

We propose integration of both models—the proposed model of the learning process as an outcome of the case study disasters, as discussed in Chapter 1 (learning from failures relies on three principles: (1) feedback to design, (2) use of advanced techniques for analysis, and (3) extraction of interdisciplinary generic lessons), and the generic lessons learnt from failures, as illustrated in Table 13.1—into a unified framework as shown in Figure 13.2.

13.3 BEST PRACTICE OF LEARNING FROM FAILURES FROM DIFFERENT INDUSTRIES

Flying is commonly considered to be one of the safest ways to travel. For example, according to Dijkstra (2006), fewer people died in commercial airplane accidents in the United States over the past 60 years than are killed in the US auto accidents in a typical 3-month period. The International Civil Aviation Organization (ICAO) stresses in its standards (Annex 13) that the intention of an investigation into an aviation accident is not to put blame on people but to create a learning opportunity. So a typical investigation in that industry will result in a report with safety recommendations aimed at preventing subsequent accidents by improving, changing, or implementing new safety barriers (Dijkstra, 2006).

13.4 **BEST PRACTICE CAN BE LEARNED FROM WORST PRACTICE**

According to Elkind et al. (2011), "*Despite efforts to change, BP never corrected the underlying weakness in its safety approach, which allowed earlier calamities, such as the Texas City refinery explosion. Perhaps the most crucial culprit: an emphasis on personal safety (such as reducing slips and falls) rather than process safety (avoiding a deadly explosion). That might seem like a semantic distinction at first glance, but it had profound consequences.*" Consider this: BP had strict guidelines barring employees from carrying a cup of coffee without a lid—but no standard procedure for conducting a "negative-pressure test," a critical last step in avoiding a well blowout. If done properly, that test might have saved the *Deepwater Horizon* platform.

Chapter 14

A Model of Learning and Unlearning from Failures

14.1 **INTRODUCTION**

Unlearning has been observed in many recent disaster investigation reports. For example, the investigations of the BP/Transocean Deepwater Horizon accident (CSB, 2010; BP Commission Report, 2011; DHSG, 2011; USDOI, 2011) claim that preventive lessons should have been learned from the reports of the investigation team into the earlier BP Texas City accident (Baker, 2007). The same applies to NASA where, in the case of the Columbia accident, the Investigation Board report (CAIB, 2003) and the analysis of Vaughan (2005) drew parallels between the case of an engineer in the Challenger project who acted aggressively to try to avert an accident (Rogers, 1986), but was faced with management opposition, and a similar scenario that was a precursor to the Columbia disaster (CAIB, 2003), although in a different context. What is interesting is that in every single in-depth analysis of a major disaster reported here, we have found that the official investigations have concluded that previous lessons from accidents, or near-misses, have not been learned. In fact, research indicates that decision makers often evaluate near-misses events as successes rather than problems or failures (Dillon and Tinsley, 2008). This can be seen in the governmental report on the Piper Alpha oil platform accident, which had a major impact on the oil and gas industry, and concluded that *"lessons of previous relevant accidents had not been followed through"* (Cullen, 1990).

Perin (2005) argues that industry incident reports render *"thin descriptions"* out of *"thick descriptions,"* not only because of inabilities to conceptualize real-time data but also because of inabilities to represent findings with relevant contextual data and as narrative, thus making it difficult to make adequate sense of events. So, in terms of learning and unlearning from failures, an important question is posed: why do some

Learning from Failures. DOI: http://dx.doi.org/10.1016/B978-0-12-416727-8.00014-X

routines stay the same when we want them to change, while other routines change when we want them to stay the same? To address this, we will challenge the traditional understanding of organizational learning from major disasters in particular by taking account of the rarity of such events.

A disaster is, by definition, a rare event of high impact. Here, we shall examine this rarity aspect and argue that, at a certain level of analysis, disasters are not rare in the sense that generic lessons can be learned and embedded in the form of dynamic routines that are continuously updated using feedback mechanisms. In other words, organizations can vicariously learn from others (Levitt and March, 1988; March, 1991). For example, when analyzing Deepwater Horizon, it can be seen that it was closer to the Apollo 13 accident than to the Exxon Valdez accident (Fowler, 2010) in that it involved having to manage systems without human eyes on the scene. We will adapt the decision-making grid (DMG) model to build a theory that explains why organizational unlearning from failures occurs. We will also examine the role of feedback in adapting routines within an organization.

We will explore similarities between the two accidents (Challenger and Columbia) at NASA, and between the two accidents (Texas City and Deepwater Horizon) at BP. We will then draw upon various common themes that can be extracted from an examination of both organizations, with special emphasis on the unlearning phenomena. We will use various theoretical lenses to perform the analysis. Finally, we will propose a theory that can enhance resilience using dynamic routines. In short, we will hypothesize mechanisms, grounded in maintenance management and decision theory and also inspired by safety research, to enhance the feedback aspect of dynamic routines.

Discussion around the general topic of feedback and learning has been very well established since the groundwork of Levitt and March (1988) on organizational learning, in which they stated that mechanisms for learning from experience are still largely unspecified. The approach that will be presented here aims to provide a mechanism to facilitate such a learning process, building on three observations drawn from organization studies and safety science.

The first observation is that a recording and transmission mechanism for routines can be a suitable environment for dynamic feedback adaptive to events related to failures and near-misses (Nelson and Winter, 1982). The second observation is that risk assessment based on one dimension that relates to probability of occurrence is inadequate for providing a comprehensive view for evaluation (Heinmann, 2005). A two-dimensional view that takes into consideration impact in addition to frequency offers a rich framework for selection of appropriate response strategies, necessary for the inclusion of both human and systems approaches. The third observation is that experience gained from learning from failures can be a rich source of generic lessons in terms of longitudinal analysis within the same organization, and in terms of comparisons between similar events in different organizations.

To help clarify the nature of dynamic organizational learning from failures, we will examine its application in the maintenance management function. There are many similarities between maintenance and safety, both in general and with particular respect to routines. Maintenance deals with breakdowns of equipment, whereas safety may be dealing with disasters. Maintenance has preventive maintenance (PM) instructions which are in many ways similar to routines in safety. Maintenance has both reactive and proactive modes, whereas in safety we deal with response to disasters in a corrective mode, and enhance resilience in a proactive mode. So, maintenance in the form of models and safety in the form of learning from failures can be combined to form a theory of dynamic adaptive routines in organization learning, which will set the ground for addressing the following research questions: What are the mechanisms for feedback and change in routine dynamics? How can organizations learn and change their routines through feedback?

14.2 RESEARCH METHODOLOGY

There are five primary accident analysis types, as defined by Stellman (1998): (i) analysis and identification of where, and which types of, accidents occur; (ii) analysis with respect to monitoring developments in the incidence of accidents, which looks at factors that affect the process operation and could lead to accidents and drives the implementation of effective monitoring of preventive activities; (iii) analysis to prioritize initiatives that call for high degrees of risk measurement, which in turn involve calculating the frequency and seriousness of accidents; (iv) analysis to determine how accidents occurred and, especially, to establish both direct and underlying causes; and (v) analysis for elucidation of special areas which have otherwise attracted attention (a sort of rediscovery or control analysis).

The different approaches for treating the organizational dimension of accidents can be classified into three main types, as proposed by Le Coze (2008): research with theorizing purposes, commissions set up for investigating major accidents, and structured root cause analysis methods. Here, we will use the theoretical lens of classification such as a framework.

Accordingly, it will be argued that techniques inspired by maintenance and reliability engineering, and decision sciences, can serve as a mental model for understanding the root causes of disaster and can support the decision-making process. A theory will then be built to address the dimensions of learning from failures and will argue that using mental models for the retention of knowledge, in the form of dynamic organizational routines, can contribute to preventing, or limiting, the "unlearning" process.

The following section will present a new decision-support tool for streamlining the process of extracting generic lessons from major disasters and feeding them back, in an adaptive process, to organizational routines.

14.3 ROUTINE DYNAMICS IN LEARNING AND UNLEARNING FROM FAILURES

14.3.1 Literature on Organizational Dynamic Routines

The dynamic nature of organizational routines was proposed by Feldman (2000) who argued that they have a great potential for change where previously they have been perceived, and even defined, as unchanging. Feldman and Pentland (2003) who argued that routines are *"repetitive, recognizable patterns of interdependent actions, carried out by multiple actors,"* also argued that routines are dynamic and an important source of flexibility and change.

Organization routines have been researched in terms of "dualities," such as exploitation of new possibilities and the exploration of old certainties in organizational learning (March, 1991). They have also been researched as a duality of stability and change and compared in terms of their mechanism and outcome (Farjoun, 2010). They have been classified in terms of regularities by Salvato and Rerup (2011), as behavioral regularities (Winter, 1964) in terms of recurring analytic processes embedded in organizations and, performed by groups of individuals, in terms of cognitive regularities (March and Simon, 1958; Cyert and March, 1963), abstract patterns, or understandings that organizational agents adopt to guide and refer to specific performances of a routine.

Feldman and Pentland (2003) adapted a taxonomy for routines that was originally conceived by Latour (1991) in his analysis of power and proposed a duality that shapes routines in terms of the ostensive aspect of a routine (which relates to the principal structure), and the performative aspect (which relates to its specific contents in terms of practice and implementation). In this line of research, the idea is to integrate routines as both the ostensive and the performative are created and recreated through action.

Organizational routines have also been defined in terms of the concrete level (Pentland et al., 2012) which was referred to as the "performative" aspect of routines based on Feldman and Pentland's (2003) definition which included specific actions taken at specific times and places. At the same time, organizational routines have been defined (Pentland et al., 2012) in terms of the abstract level, referred to by Feldman and Pentland (2003) as the "ostensive aspects," i.e., the abstract patterns of the routine. It was also argued by Pentland et al. (2012) that at the macro-level, the dynamics of routines emerge from the micro-level relationship between specific actions and patterns of action.

D'Adderio (2008) has proposed a duality of *(re)framing*—in the form of prescribing actions and overflowing—in the form of interpreting rules and procedures. Levinthal and Rerup (2006) studied categorization of routines in terms of the duality of mindful and less-mindful processes, and suggested that the ostensive/performative classification is useful for understanding the two processes.

In social science, as in many walks of life, things are not usually classified as Boolean (i.e., as separate entities). There are always interactions and interdependent issues in a real, complex, situation, and hence the authors who proposed these dualities (exploitation and exploration, stability and change, behavioral and cognitive, ostensive and performative, concrete and abstract, framing and overflowing, mindful and less-mindful) have also pointed out that there is a degree of feedback, interaction, complementary relationships, and recursive relations between the components of these dualities. The same applies to knowledge management, in terms of the duality of tacit and explicit knowledge as proposed by Nonaka and Takeuchi (1995) and the transformation process, among and between them.

Despite the recognition of interaction among the foregoing dualities, such interactions had not always been theorized at the time they were realized. For example, Miller et al. (2012) have argued that the reciprocal relation between the ostensive (the abstract pattern) and the performative (the specific instantiations), as proposed by Feldman and Pentland (2003), did not elaborate on the mechanisms by which these aspects interact. Miller et al. (2012) have proposed to fill this gap by translating agents' memory into performances and using simulation to highlight the importance of transactive memory. They base their research on three types of memory—procedural (know-how), declarative (know-what), and transactive (know-who) to explain routine dynamics. D'Adderio (2008) studied interactions between aspects of routines in the form of influence of artifacts, such as standard operating procedures (SOPs) for routines for freezing software design and their impact on performance.

In the maintenance domain, research suggests that PM instructions should be dynamic and adaptive to the shop-floor. They are generated (initiation of new routines), older and irrelevant ones needing to be dropped after careful consideration (discarding nonrelevant routines), and existing sets of PMs needing to be prioritized (ranking of existing routines).

The initiation of new routines can be an outcome of machine failures and based on update of the equipment manual, suggestions by the vendor, or on new improvements or modifications of the design. The main challenge is that once new instructions are generated, the database of these will expand, and in the long term such a list will be neither effective nor efficient for the stability of the organization. Hence, there is a need to balance this expansion with deletion of nonrelevant routines, which should be carefully managed because disasters, by their very nature, are rare and a particular routine might not be used frequently but will be needed in the event of a disaster.

For example, in the case of the Bhopal disaster, the MIC storage 30 tons refrigeration system was installed to keep the storage tank material, which was hazardous, below 5°C. However, the refrigeration system had been shut down in May 1984 to save power, a saving of approximately $20 US/day. This has been cited as one of the reasons for the disaster in December of the same year, and the loss of more than 3000 lives (Chouhan, 2005). This instruction (routine) to use the refrigeration system was in the original designer's manual of routines, but it had not been perceived as relevant by the management because they wrongly assumed that, since it had not been used before, it could be discarded.

The above is an example of a dilemma with respect to the dynamic nature of routines. On the one hand, to improve safety we encourage creation of new ones, while on the other hand to raise efficiency and minimize costs management we try to discard existing ones. Decisions to downgrade the criticality of routines and instructions need to be carefully challenged.

The third aspect of dynamic routines with respect to instructions is that the existing list of routines needs to be continuously reprioritized in the light of problems and near-misses, changes in the environment or design, and risk assessments. This process of ostensive dynamic adaptive routines is more easily talked about than done. In the next sections, a new theory is provided to help with dealing with such a challenge.

14.3.2 **Literature Related to Learning from Disasters**

Research indicates that organizations learn more effectively from failures than from successes (Madsen and Desai, 2010), that failures contain valuable information but organizations vary at learning from them (Desai, 2010), and that organizations vicariously learn from failures and near-failures of others (March, 1991; Kim and Miner, 2007; Madsen, 2009).

It is generally accepted, however, that learning from failures is a difficult process to assess. Few authors have attempted to define it. Organizational learning from failures has been defined by Madsen and Desai (2010) as any modification of an organization's knowledge occurring as a result of its experience. But again it is acknowledged that a change in organizational knowledge is itself difficult to observe. Subsequently, there has been a trend in research which argues that changes in observable organizational performance reflect changes in organizational knowledge (Argote, 1999; Baum and Dahlin, 2007). Another line of research has tried to explore ways of learning. For example, Carroll et al. (2002) proposed a four-stage model of organizational learning that reflected different approaches to control and learning. Also, Chuang et al. (2007) developed a model of learning from failure of healthcare organizations that highlighted facilitating factors.

Organizations learn when individual knowledge is codified, synthesized, and transformed into new technologies, training programs, policy, regulations, plans, and organizational structure (Zollo and Winter, 2002). Organizational learning can happen not only across hierarchical levels of the organization through transformation between tacit and explicit knowledge (Nonaka and Takeuchi, 1995) but can also occur within organizational routines (Feldman, 2000). Pentland et al. (2012) argue that routines have been theorized as a primary mechanism for organizational learning and, citing Levitt and March (1988), they argue that routines tend to improve over time, at least in the early stages of formation (Argote and Epple, 1990; Narduzzo et al., 2000; Zollo and Winter, 2002; Rerup and Feldman, 2011).

Rare events have traditionally been on the margin of mainstream research in organizational learning, as they are often set aside as mere statistical outliers (Lampel et al., 2009). However, a special issue of *Organization Science* in 2009 was dedicated to rare events and organizational learning. The editors (Lampel et al., 2009) proposed a taxonomy of learning based on two categories: the impact of rare events on the organization and their relevance to the organization. Based on those, they identified four classes of events: transformative, reinterpretative, focusing, and transitory. They then mapped the works of Beck and Plowman (2009), Christianson et al.

(2009), Madsen (2009), Rerup (2009), Starbuck (2009), and Zollo (2009) into those classes. Using this theoretical lens, this chapter will examine disasters that have had high impact and hence may be regarded as either transforming or focusing depending on their degree of relevance to the specific organization. However, there will also be a review of generic lessons that are applicable to a wide range of industries.

Lampel et al. (2009) argued that further research is needed on how organizations can learn and adapt their routines, following rare events in other organizations. They argued that the aviation industry is a good example, in which the lessons from accidents, near accidents, and the possibility of accidents are codified into rules and practices that change how airlines operate (Tamuz, 2000; Lampel, 2006).

There is evidence of a lack of research into how the learning process can emerge or how to use models that can facilitate the process of learning from failures and extracting generic lessons. More specifically, there is a lack of research into how this learning can be dynamically fed back into organizational routines.

In 2005, an overflow of petroleum fuel at the storage plant at Buncefield, in the United Kingdom, and the subsequent ignition of a vapor cloud, caused a massive explosion and fire. An investigation by the United Kingdom's Strategic Management Group found that the safety management system focused too closely on personal safety and that there was a lack of in-depth analysis of the control of major hazards (Buncefield Major Incident Investigation Board, 2008). In the same year, the BP Texas City refinery experienced a disaster when uncontrolled release of a flammable liquid led to an explosion and major fire. The subsequent reports concluded that *"BP emphasized personal safety, but not process safety"* (BP Texas City Report, 2007). It is this *"tick the box"* mentality and an observation that the organization *"forgot to be afraid,"* which is a recipe for unlearning and drift toward another disaster. Five years after the Texas City disaster, we had the Deepwater Horizon event. The key issue about learning is not just admitting that learning has occurred but how this has been fed back to routines within the organization in a continuous adaptive way.

14.3.3 **Learning and Unlearning from Disasters—The Case of NASA**

It is interesting to note the striking similarities between the Challenger and Columbia disasters. Mahler and Casamayou (2009) suggest that, given the similarities in their causes, there may be evidence that NASA is *"not a learning organization."*

According to Vaughan (1996, 2005), both disasters were due to technical issues (O-ring blow-by in the case of Challenger, foam shedding in the case of Columbia), each of which had been experienced many times previously, but which had been "normalized," although individuals in the organization had expressed concerns. There were organizational problems highlighted by both investigative commissions and yet, according to Leveson et al. (2009), the same organizational problems that created the Challenger accident were implicated in the Columbia accident.

The main generic lesson that can be learnt from the Columbia disaster is perhaps about the concept of "unlearning." According to Mahler (2009), there are three categories of lessons that organizations can derive from disasters: (1) not learned, (2) learned only superficially, or (3) learned and then subsequently unlearned. The idea of learning and then unlearning is of particular interest as the concept of unlearning has received relatively little attention in the literature. An example of the "unlearning" at NASA concerns its organizational structure, where according to the Rogers report (1986), among the reasons for the Challenger disaster were the existence of rivalry among centers, different organizational cultures, lack of responsibility and accountability, and cross-center communication issues which created information and coordination problems. After the disaster, the shuttle program was restructured into a centralized configuration which continued until the 1990s, but then reorganization occurred and the system was decentralized, to improve efficiencies; a policy of "faster, better, and cheaper" according to Woods (2006). This policy of coping with pressures created the conditions favoring the Columbia disaster (CAIB, 2003).

With respect to the organization structure and routines involved, there are some basic generic principles important in providing an effective structure for organizing system safety, as proposed by Leveson (1995): (1) System safety needs a direct link to, and influence on, decision makers (influence and prestige). (2) System safety needs independence from project management (but not engineering). (3) System safety needs direct channels of communication to most parts of the organization (oversight and communication).

Evidence was revealed, after the Columbia disaster, of failure to satisfy the above-mentioned objectives. For example, in terms of influence and prestige, the CAIB (2003) report concluded that system safety engineers were often stigmatized, ignored, and sometimes actively ostracized. It states that "*Safety and mission assurance personnel have been eliminated [and] careers in safety have lost organizational prestige.*" In terms of independence, the report noted that "*Given that the entire safety and mission assurance organization depends on the Shuttle Program for resources and simultaneously lacks the independent ability to conduct detailed analyses, cost and schedule pressures can easily and unintentionally influence safety deliberations.*" Finally in terms of oversight and communication, it was found that "*organizational barriers ... prevented effective communication of critical safety information and stifled professional differences of opinion.*" It was "*difficult for minority and dissenting opinions to percolate up through the agency's hierarchy.*"

NASA's unlearning from failures was evidenced by its staff cuts to cope with budgetary pressures, which ultimately led to the disappearance of organizational memory. Kletz (2009) claimed that people have memories whereas organizations don't. It is therefore important to develop innovative mental models that can contribute toward knowledge retention within organizations.

14.3.4 **The Case of BP**

There were striking similarities between BP's Texas City and Deepwater Horizon disasters. According to the investigation into Deepwater Horizon, it had eerie similarities to Texas City (DHSG, 2011). Among these that specifically related to organization routines were: (a) not following required or accepted operational guidelines "*casual compliance*"—at Texas City proper guidelines for the level of liquid in the tower were not adhered to, whereas at Deepwater Horizon proper cementing procedures were not followed—a case of "*normalization of deviance*," a term coined by Vaughan (1996); (b) neglected maintenance routines related to instrumentation that either did not work properly or whose data interpretation gave false positives; (c) inappropriate assessment routines and management of operations risks; and (d) lack of appropriate selection and training routines of personnel.

In both disasters, worker safety and systems safety were confused, and meetings were held with operations personnel at the same time and place as and where the initial failures were developing. These meetings were intended to congratulate the operating crews and organizations for their excellent records for worker safety. According to Vinnem (2013), both of these disasters have served to clearly show there are important differences between worker safety and system safety. One does not assure the other.

Both these cases of unlearning—from NASA and BP—demonstrate the need to investigate why organizations and institutions fail to learn, and what the mechanisms are for feedback and change in routine dynamics. In short, how can organizations learn and change their routines through feedback?

14.4 A NEW THEORY OF ROUTINES FOR ADAPTIVE ORGANIZATIONAL LEARNING FROM FAILURES

To a normal person, a fire may be a major accident, or a disaster, while to a fireman it will be just a daily routine job. It is argued here that although disasters are rare events of high impact, an approach can be developed to make it a routine to analyze them and to establish a dynamic feedback mechanism for facilitating such a scheme.

On the other hand, to a normal person a repetitive and recognizable pattern can be looked at as a routine, whereas for those implementing it there may be a high degree of dynamic and creative processes involved in carrying it out. For example, consider swimming or running as a daily activity. It can be perceived as just another routine, but to a professional swimmer, every day is different in setting new targets for distance and speed, developing new styles of breathing and stretching, swimming strokes, etc. Another good example is academic recruitment as a routine, which was used by Feldman and Pentland (2003) to illustrate the dynamic nature of routines in terms of being ostensive or performative. Again, it is argued here that by analyzing frequent events of low significance, such as near-misses, the theory can be enhanced to make it a routine to analyze such cases and to establish a dynamic feedback mechanism. It is also argued that for each dimension of frequency and significance, and on a scale of low to high extremes, there is a middle range, where routines in the form of preventive measures (similar to routine checks in PM instructions) can be adopted to help the organization to become more resilient.

14.4.1 Lessons Learnt from the Maintenance and Reliability Field

It is widely acknowledged that safety is created through proactive resilient processes, rather than through reactive barriers and defenses (Wood and Hollnagel, 2006). The transformation from reactive to proactive modes in maintenance and reliability can be achieved, broadly speaking, via two schools of thought: human oriented and system oriented. In short, maintenance policies can be broadly categorized into the technology or systems oriented (systems or engineering) on the one hand, and the management of human factors oriented, and monitoring and inspection oriented on the other (Labib, 2008).

14.4.1.1 *Human-Oriented Approach Based on the TPM Concept*

Total productive maintenance (TPM) is a human-factors-based concept in which maintainability is emphasized. It originated from the Japanese manufacturing systems (Nakajima, 1988; Hartmann, 1992; Willmott, 1994), and is a tried and tested way of cutting waste, saving money, and making factories better places in which to work. It gives operators the knowledge and confidence to manage their own machines. Instead of waiting for a breakdown, and then calling the maintenance engineer, they deal directly with small problems, before they become big ones. Operators investigate and then eliminate the causes of minor and repetitive machine errors. Also, they work in small teams to achieve continuous improvements in the production lines.

One of the underpinning elements of TPM is the skill levels needed and the transformation of some of the basic maintenance skills from maintenance engineers to the frontline operators in production, and hence the term "productive maintenance." The designer of the machine is not usually the one who, on its failure, fixes it and, surprisingly, might not even have the ability to do so. For example, skills needed to restore equipment may include diagnostics, logical fault finding, disassembly, repair, and assembly. Depending on the level of complexity of the particular equipment, as well as on the level of complexity of the function that it carries out, the necessary skill level can be determined.

According to a survey conducted by McDonald (2006) of aircraft maintenance technicians, they reported that in approximately one-third of the tasks they did not follow the routine procedure according to the maintenance manual. They felt that there were better, quicker, and even safer ways of doing the task than following the manual to the letter. McDonald (2006) argued that manuals themselves are not an optimum guide to task performance as they have to fulfill other criteria, such as being comprehensive and up-to-date. The question is: How to bring operator requirements to the forefront of the design process? Or how to feedback the knowledge, skills, and experience of the operator who is day in day out in front of the machine, to the designer? In a crisis, the skill levels and types needed to constitute a major dilemma because disasters tend to be multidisciplinary problems. They can span various fields such as information systems, maintenance, decision making, and crisis and risk management, and hence there is a need for a synchronized multidisciplinary team approach.

14.4.1.2 *Systems-Oriented Approach Based on the RCM Concept*

Reliability centered maintenance (RCM), originated in the aviation industry, is a system and technologically based concept in which the reliability of machines is emphasized. The name was originally coined by Nolan and Heap (1979). It is a method for determining the maintenance strategy in a coherent, systematic, and logical manner (Moubray, 1991; Netherton, 2000). It offers a structured procedure for determining the maintenance requirements of any physical asset in its operational context, the primary objective being to preserve system function. The process consists of looking at the way equipment fails, assessing the consequences of each failure (for production, safety, etc.), and choosing the correct maintenance action to ensure that the desired overall level of plant performance (i.e., availability, reliability) will be met.

One of the underpinning elements of RCM is the routine of root cause analysis. When performing reliability analysis investigation using tools such as FTA, the main aim is to identify the root causes (i.e., at the bottom of the tree). These basic events are considered to be the "leaves" of the tree, the initiators or "root causes." Here the term "root cause" needs to be treated with care and it is also important to differentiate between the concept of root cause for machines or equipment as compared with that for a disaster or an accident.

In an accident investigation, if root cause is perceived as, for example, someone's behavior then it may be likely, as argued by Rasmuseen (1997), that the accident could occur by another cause at another time. We agree that, in this example, such a root cause is superficial and should be regarded as still part of the symptom rather than the real root cause. A real root cause needs to be plan and policy related with respect to the current status quo. As such, ideally a root cause should lead to initiation or modification of a routine in the form of SOPs. When digging deeper to find a real root cause, the answer is the keyword phrase of "lack of procedures" to do a certain task. We are looking for a status quo that is not acceptable, a lack of existing standard. Hence resolving this issue (through establishing a new routine that is currently missing) will lead to establishing a generic solution that will solve many similar problems.

14.4.2 **The DMG**

Here, we shall revise and extend the DMG originated by Labib (1998, 2004) which has been tested by organizations for the selection of appropriate maintenance strategies for machines on the shop-floor of a manufacturing environment or process industry (Fernandez et al., 2003; Burhanuddin, 2007; Tahir et al., 2008; Zainudeen and Labib, 2011). The DMG acts as a map in which the worst machines are represented according to multiple criteria of their performance. The objective is to identify appropriate actions that will lead to the movement of the machine location, on the grid, to positions indicating improved states with respect to the multiple criteria of performance in terms of frequency and downtime.

The revised model proposed here will be different from this in two ways: it will be applied to learning from disasters, and it will address the incorporation of organizational routines. The scale of the frequency axis will be based on the rate of failure, or on the incidence of chronic problems, whereas the significance axis will be based on a measure of the acuteness of the incident (i.e., of its severity, cost, etc.).

Events occurring on any asset (e.g., equipment or artifact) are plotted with respect to their relative performance, i.e., as regards their frequency and significance (Figure 14.1). The objective is then to implement appropriate strategies that will lead the movement of the representation of each asset toward location on the grid indicating improved status.

The procedure is as follows:

i. *Criteria analysis*: Establish Pareto analysis of the criteria, in terms of relative significance and frequency. For a disaster, these would be expressed as "acute-ness" and "chronic-ness" as opposed to the "downtime" and "frequency" of the traditional DMG.
ii. *Decision mapping*: Map the criteria on the matrix, i.e., plot each event for any asset on the grid.
iii. *Decision support*: Identify an action, based on the suggested strategy, to be implemented.

The next section will explain the suggested safety measures in the case of learning from failures and the impact of this on routines.

14.4.2.1 *Suggested Strategies*

Here, we will first address the "extreme case" strategies, i.e., those indicated in Boxes 1, 3, 7, and 9 of Figure 14.1 (i.e., at the corners of the grid), and then those indicated in Boxes 2, 4, 5, 6, and 8 (the other less extreme cases).

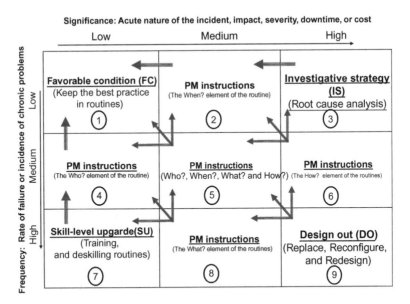

Figure 14.1 Frequency/significance grid for deciding routines to meet threats of a given incident.

1. *Favorable condition (FC)*: Box 1, low significance and frequency. Compared to others, the asset fails rarely, and failure is insignificant, i.e., only just noticeable. We would wish all events to fall in this category and the aim is to suggest appropriate strategies for the events indicated in the other boxes so that their indications would move toward this box. In the safety domain, this region is about sustaining current practice, i.e., business as usual, and hence keeping the status quo with respect to the routines for the assets concerned.

2. *Investigative strategy (IS)*: Box 3, breakdowns are infrequent but restoration of the asset takes a relatively long time. In the maintenance function, the asset is perceived as problematic "a killer" (Labib, 2004); in the safety field, a high significance event but a rare one, which can feature in a disaster. Whenever an airliner crash occurs, it is of extreme severity (incurring deaths and the loss of the plane). But nobody ever heard of the same airplane (asset) crashing twice: a case of extreme rarity. In such an eventuality, search teams look for the black box flight recorder, which monitors, as much as possible, every detail of the flight prior to the crash. The strategy investigative, the focus being to find the root cause and make recommendations (new routines) in order to either eliminate the possibility of reoccurrence of such a crash or to mitigate the impact

of such an event. The idea of using the black box device—in many ways similar to condition-based maintenance (CBM), which relies on measuring vibration, acoustic emissions, etc—is to maximize access to data, in a situation which, due to the rarity of its occurrence, offers minimal opportunity for collecting information and hence for an investigative approach.

3. *Skill-level upgrade (SU)*: Box 7. In the maintenance field, this could refer to a task on a machine that is visited many times (high frequency), but for limited periods because the task involved is easy to complete (low downtime). Upgrading the skill level of the operator is key, so that this relatively easy task can be reliably assigned to the operators after upgrading their skill levels. Another variation of this strategy is to upgrade the machine so that its maintenance (the diagnosis or preventative measures required) is "deskilled." For example, in modern paper photocopier machines, one of the relatively frequent events, which is also relatively simple to fix, is the paper jamming inside the machine. Hence, we make it easier for a normal person (operator) to fix it, rather than calling a maintenance engineer, by installing a display that shows the structure of the photocopier and the location of the jammed paper. Also, when the doors of the photocopier are opened, we note that the knobs and levers that can be accessed are colored in an attractive green or blue, making it easier to identify which parts to touch or not. Then, when the job is finished, the doors are designed in such a way that if a certain lever is left in the wrong position, the doors will not shut. Again, the idea is to "deskill" the routines needed to respond to such frequent nonsignificant problems. In the safety domain, we are here addressing frequent incidents of low impact and the suggestion is to focus on training and on raising awareness of how to implement, in the most efficient way, safety measures, and SOPs as routine, either by upgrading the skills of the operators involved or by deskilling the routines needed to respond to such incidents. In the safety domain, a good example is the near-miss, a situation where a bad outcome could have occurred except for the fortunate intervention of chance (Dillon et al., 2012).

4. *Design-out (DO)*: Box 9, the most crucial area in the grid. In maintenance, machines in this category are recommended for major (DO or overhaul projects. This is because they experience high downtime with high frequency. In the safety field, we are dealing here with a disastrous situation that has been repeated, e.g., NASA's Challenger and Columbia, and BP's Texas City and Deepwater Horizon events. The strategy is to examine a situation that is

currently not fit for purpose and to reconfigure or redesign it. Either terminate the status quo (e.g., stop the space shuttle program) or adopt a resilience approach, with a fundamental reconfiguration of organizational structure with the emphasis on preventing reoccurrence, minimizing its impact, and increasing ability to detect and monitor (a combination of strategies suggested in Boxes 3 and 7). Note here that both Boxes 3 and 9 indicate the same degree of significance, but the strategies suggested are very different, because they need two different mind-sets for framing the problem. This aspect, of mind-set, will be described in Section 14.6.

5. *PM instructions routines*: Boxes 2, 4, 5, 6, and 8. If one of the antecedent events is of medium significance or medium frequency, the suggestion is to focus on modifying the current preventive routines and on their nature. Consider, for example, a simple PM routine with respect to a car. It might be as follows: *"Change the oil filter, the mechanic to use a certain spanner, every six months or every six thousand miles, whichever comes first."* This simple instruction contains several features, i.e., *when* or how often it takes place (every 6 months or every 6000 miles), *who* will carry it out (the mechanic as opposed to the driver), *how* it will be done (using a certain spanner), and *what* is the nature of the instruction itself (change the oil filter). Hence it raises "when?," "who?," "what?," and "how?" types of questions.

The adaptive feature of *ostensive* routines contains three issues: the need to generate new routines, the need to get rid of noneffective ones, and the need to prioritize existing ones, whereas the adaptive feature of *performative* routines contains two main issues: regarding the relative "easy" aspects of the PM instructions, such as who will perform the routine (a skill-related issue) and when it will be performed (a timing or schedule-related issue), and relatively "difficult" aspects such as what is the nature or content of the instruction and how will the instruction be performed? In order to appreciate why the former issues are relatively easy, consider the case of giving a drug to a patient to prevent a disease as an analogy to applying a PM instruction/routine in maintaining certain equipment. The easy issues/questions concern who will administer the drug and the suggested dose, whereas the relatively difficult issues relate to questioning whether it is the right drug in the first place (a diagnostic expert skill is needed here) and how to perform the instruction in the right way (again a matter of high expertise). In the safety field, this is the region where we analyze our existing PM routines, basic checks of systems, safety barriers, and backup systems.

However, not all of the boxes that indicate a medium component are the same as we shall now see:

1. *Easy aspect of the PM routines*: This is concerned with the "*who*" and "*when*" type questions of the routine: There are some regions, such as those indicated by Boxes 2 and 4, that are near to the favorable top left corner of the grid (Box 1), which are concerned with the relatively "easy" aspects of preventive routines. Hence, they require readdressing issues regarding "*who*" will perform the instruction or "*when*" the instruction will be implemented. For example, in some cases, the issue may be about who will carry out the instruction—operator, maintenance engineer, or subcontractor—which would suggest that this applies to Box 4 situation, that box having a border with Box 7 that relates to the type of skills needed (a "*who*" question), and the aspect of the PM instruction is relatively "easy" to implement, which normally denotes either a "*who?*" or a "*when?*" type of question. The "*when*" type question in the routine is also based on the same line of argument; if the event is located in Box 2 due to its relatively higher significance (or in maintenance words its higher "downtime"), then the timing of instructions needs to be addressed which is a "*when?*" type of question with respect to the preventive routine.

2. *Difficult aspect of the PM routines*: Other preventive routines, such as the ones related to Boxes 6 and 8, need to be addressed in a different manner. Again, the "difficult" issues here are those related to the contents of the instruction itself. It might be the case that the wrong problem is being solved, or that the right one is not being solved adequately. In other words, we are giving the patient a drug at the right time and in the right dose, but unfortunately it is not the right drug in the first place. In this case, routines need to be investigated in terms of the content of their instructions, and expert advice is needed regarding the type of routines being implemented, the "*what?*" and "*how?*" types of question. These two types of "difficult" question apply to Boxes 6 and 8 since they have borders with the "worst" box Number 9. For a Box 6 routine, the issue concerns the "*how*" aspects, as the time factor being the main problem in this case. As for a Box 8 routine, the issue is the "*what*" aspects, since the frequency of the event is high, which indicates that the routine in question is not fit for purpose. Figure 14.2 indicates the "easy" and "difficult" parts of the routines.

Figure 14.1 also indicates that the human-oriented approach based on the TPM concept tends to occupy the bottom left triangle of the diagram,

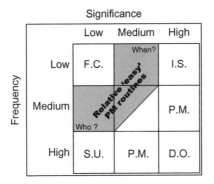
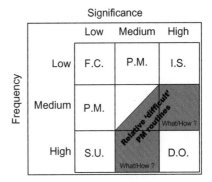

■ **Figure 14.2** "Easy" and "difficult" aspects of PM routines.

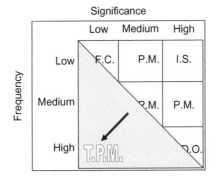
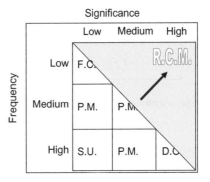

■ **Figure 14.3** The TPM (human-factors-oriented) approach versus the RCM (systems-oriented) approach.

whereas the systems-oriented approach, based on the RCM concept, tends to occupy the top right triangle, as further illustrated in Figure 14.3.

The main contribution of the proposed frequency/significance grid is its provision of a classification of suggested strategies for different aspects of routines, depending on the relative performance of events as indicated by it. Thus, it represents a dynamic feedback approach to routines in a dynamic environment.

14.5 CASE STUDY OF APPLYING THE PROPOSED MODEL TO A DISASTER ANALYSIS

In one near-miss event, NASA's space shuttle, despite having undergone a DO program, had to be repaired during a mission, on the discovery of a

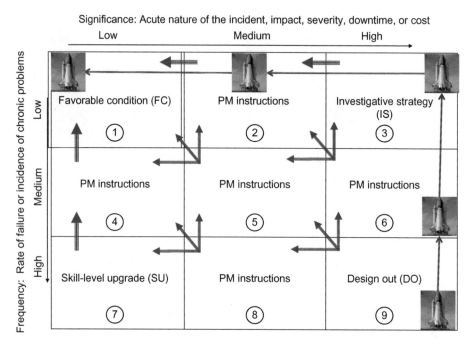

■ **Figure 14.4** The frequency/significance routine-selection grid, applied to the NASA case.

misplaced tile, by an astronaut with the help of a robotic arm. The damage could have resulted in a repeat of a catastrophic failure. The question is how can one use the proposed model to map the different stages through which the shuttle had gone, starting from the DO stage?

The basis of a possible solution is shown in Figure 14.4.

The shuttle was redesigned—(a) DO—after two disasters that had occurred in previous missions, firstly the Challenger and then Columbia. In particular, with respect to the Columbia disaster, one of the root causes found was lack of imagery information for assessing the significance of damage on the wings after foam had hit tiles on it. The DO had resulted in the installation of cameras and remote monitoring devices (a CBM strategy or, in general terms, an investigative strategy (IS)), that were able to capture the disoriented part. When the problem happened again, the reduction of significance, or severity, due to the CBM-prompted astronaut repair enabling the shuttle to land safely, put the event into Low Frequency Low Severity range, the situation having been returned to normal and a repeat disaster avoided. In summary, this study shows how the condition of the shuttle had been moved via DO from a Box 9 state to a Box 6 state (improving the routines) to a Box 3 state (implementing an IS), and then to a Box 2 state (enabling the routine of assessment and repair in real time), and ultimately to a Box 1 (FC) state. In other cases, the outcome of the DO might lead to moving to an SU and, eventually, a favorable FC state.

14.6 THEORETICAL FRAMEWORK AND DISCUSSION
14.6.1 Near-Misses and Failures

One of the most critical steps in the learning process is the classification of prior experience as success or failure (March, 1991). Near-misses are, however, often framed as success (Dillon and Tinsley, 2008). It can be argued that deciding whether near-misses are successes or failures depends on the way people interpret the meaning of the word "failure." According to the British Standard BS4778 and the American Military Standard MIL-STD-1629A, 1980, failure is the *"termination of the ability of an item to perform its function."* The word "termination" implies that a machine has two states (working or not), i.e., the transition from uptime to downtime is sudden. The element of deterioration is not captured in this definition, so the issue is whether we define a failure as something that does not work or is unavailable (i.e., termination of its ability to perform its function), or rather as something that is not operational at 100% performance (which suggests deterioration). Thus, in a near-miss situation, any drift from optimal capability should be captured as an event that warrants analysis. Technologies such as CBM allow for such high-level capture of deterioration in performance (Holroyd, 2000; Jardine et al., 2006). It has been suggested that capturing near-misses can be enhanced by elevating the safety message throughout the organization and by emphasizing the significance of the projects (Dillon et al., 2012). While we can agree with the idea of elevating the safety message, it is unclear how elevating the significance of a project can be put into practice. For example, in the field of maintenance, it is not possible to give a high criticality score to all machines, because criticality would then lose its significance.

14.6.2 High Severity with Low Frequency Versus High Severity with High Frequency

During a semiinformal interview, an oncology surgeon stated that *"It is well known—in general—that there are at least two types, or phases, of tumors. The first type is localized and such tumors are usually treated through surgical intervention. The patient then spends a period of convalescence before recovering and may need some rehab for a limited period according to tumor size and location, and rarely needs a short period of chemotherapy or radiation to ensure that that the tumor does not return. Then the patient is subject to a precise follow-up for a period of time, for the purpose of early detection and eradication of the tumor if it appears again. The second type, and the most dangerous, is the type that does not only spread in one place (organ) but transfers to various organs of the body through the blood and lymph nodes, and this process is called 'metastasis,' where other infected organs behave like the original tumor and can sometimes be more aggressive than the original one depending on their place, in the brain or lung or joints, bones or liver. Treatment of this second type is more difficult and dangerous, and it needs patience and persistence, where surgery is not the main solution—but may be part of it and in some cases may cause greater destruction—but the treatment here is primarily through chemotherapy and radiation. The aim here is the elimination of these cells which are circulating in the whole body, dormant or active. In the light of this war, the body weakens and sometimes suffers. The treatment affects healthy cells also, sometimes before it eliminates the infected. It is a battle of will and steadfastness and the possibility of healing and hope before being just a challenge of a drug treatment. Sometimes, healing is the result; in this case vigilance and more accurate follow-up and commitment are needed. We medical doctors have been taught that a skilled surgeon is the one who knows when NOT to operate."*

In the frequency/significance grid, there are two corners of high impact (a cancer for the organization); a disaster is a rare event of high impact (low frequency but high impact), which is equivalent to the first type of cancer. Repeated disasters are of high impact and high frequency and hence they resemble the second type of cancer. It is the type of cancer that manifests itself in the "disability to learn from failures" and in the existence of a safety culture that is not fit for purpose. Surgical treatment is no longer the solution. Unfortunately, the organization involved has left this tumor spreading in the body under different names and allegations. It is not just an incident that occurred at a specific time and that demands a cure in the short term.

14.6.3 **Holonic Theory**

The concept of a holon originated from Koestler (1989) who described it as the hybrid nature of subwhole/parts in real life systems. The word comes from "holos" or "whole" which means something that is simultaneously a whole and a part. The holonic concept was originally introduced by Koestler in his analysis of hierarchies and stable intermediate forms, in both living organisms and social organizations. From this perspective, holons exist simultaneously as self-contained wholes in relation to their subordinate parts and as dependent parts when considered from the inverse direction. In other words, holons are autonomous, self-reliant, units that possess a degree of independence and handle contingencies without asking higher authorities for instructions. But at the same time, they are simultaneously subject to control from one or more of these higher authorities. The first property ensures that they are stable forms that are able to withstand disturbances, while the latter property signifies that they are intermediate forms, providing a context for the proper functionality for the larger whole, a combination of flexibility and stability.

It is argued here that disasters happen when a breakdown in control, communication, and prioritization of routines occurs. Using the holonic concept, when information flow in command and feedback, and the understanding of roles, are compromised, for any reason, the system begins to break down. In other words, the whole organizations no longer recognize their dependence on their *subsidiary* parts, and the parts no longer recognize the organizing authority of the whole organizations. Cancer may be understood as such a breakdown in the biological field. Communication, prioritization, and responsibility were argued by Labib and Read (2013) as the underpinning factors for learning from major disasters.

14.6.4 **Adaptive Ostensive and Performative Routines**

Given that the ostensive aspect of a routine relates to the principal structure, and that the performative aspect relates to its specific contents in terms of practice and implementation, one can consider events in terms of their frequency and significance and accordingly adapt their related routines as an outcome of such events. Frequency can be measured by a rate of failure, or of incidence of chronic problems, whereas significance can be measured by the acuteness of failure or problem, its impact, severity, downtime, or cost.

The ostensive and performative natures of dynamic feedback routines with respect to the generation, prioritization, and discarding of PM instructions can be considered as a holon where the nature of the PM instruction itself has an ostensive feature because it attempts to answer the "what?" and "how?" questions of the PM, and a perfomative aspect in defining the style of implementation, the timing of the routine, and the type of skill needed (Figure 14.5).

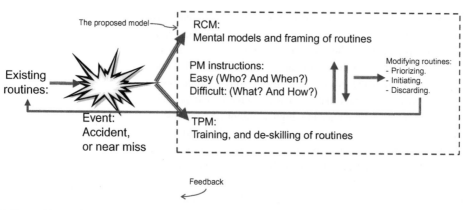

■ **Figure 14.5** Theoretical framework.

14.7 **CONCLUSION**

Focusing on learning from a specific disaster could be dangerous, because it could divert attention from other problems in the business. Moreover, simply learning as a reaction to the occurrence of just one disaster implies a passive—reactive mood, i.e., responding to "*lessons learned in blood*" (Madsen, 2009). Even worse, there is evidence that unlearning occurs after major disasters. In this chapter, it has been shown how different events can be analyzed, starting from minor incidents and near-misses through to disasters using a combination of frequency and significance. In doing so, we have identified nine strategies of management related to the dynamic feedback, into routines, of information from these events.

The contributions of this study are both theoretical and methodological. On the theoretical side, it has drawn from research into the three disciplines of organizational routines, safety, and maintenance. In the case of organizational routines, it has covered aspects of dualities in characterizing them. With respect to safety, it has addressed the different theories and demonstrated gaps in knowledge regarding learning from failures. In terms of maintenance, it has addressed the human and systems approaches and the formulation of different maintenance strategies. On the methodological side, it has combined the three fields and proposed a model of how to dynamically feedback information from events into routines.

The study also presented a decision-support grid that can classify organizational routines. There are often thousands of assets in any large organization that may fail, with different levels of severity and significance, and each of these assets and failures modes relates to many PM routines and hence in such a complex environment, that the use of the proposed classification mechanism is invaluable. Also provided is a theoretical framework for addressing near-misses, as well as the ostensive and performative aspects of preventive routines and it is argued that they are holonic in nature. In short, this study has aimed to show how to move from automatic defensive routines to automatic learning routines.

Not Just Rearranging the Deckchairs on the Titanic: Learning from Failures Through Risk and Reliability Analysis*

Ashraf Labib and Martin Read

Department of Strategy and Business Systems, University of Portsmouth, Portsmouth Business School, Richmond Building, Portland Street, Portsmouth PO1 3DE, United Kingdom E-mail: Ashraf.Labib@port.ac.uk; Martin.Read@port.ac.uk

Abstract:

How do we learn from failures? and Why do we need to learn from failures? These seemingly simple questions are yet quite profound and we need to dig deeper to explore suitable answers. In this paper an empirical study based on the analysis of reported disasters is provided with the aim of exploring techniques that can help us to understand the root causes of why those incidents occurred and how such crises unfold over time and hence how can we learn generic lessons from those disasters.

Introduction:

It has been recently reported that organizations learn more effectively from failures than from successes (Madsen, and Desai 2010) and that failures contain valuable information, but organizations vary at learning from them (Desai, 2010). It is also argued that organizations vicariously learn from failures and near-failures of others (Kim and Miner 2007), and (Madsen, 2009).

---

*Paper published in Journal of Safety Science (printed with permission) Labib, A., and Read, M., Not Just Rearranging the Deckchairs on the Titanic: Learning from Failures through Risk and Reliability Analysis, Safety Science, Vol 51, Issue 1, pp 397-413, 2013.

However, it can also be argued that lessons gained from major failures have not really been learnt by the very same organizations involved in those disasters. This is evidenced by recent reported case incidents of major organizations such as BP, NASA, and Toyota which are examples of this trend.

The first case is by BP, where in March 2005, a series of explosions and fires occurred at BP's Texas City refinery killing 15 people and injured 170 (Vaidogas *et al.*, 2008). Analysis of BP track record by (Khan and Amyotte 2007) shows that the March, 2005 disaster was not an isolated incident and concluded that BP leads the US refining industry in terms of deaths over the last decade. Yet, in April, 2010 BP suffered from an explosion that destroyed the Deepwater Horizon rig, killing 11 workers and initiated an oil spill that has not yet been rectified at the time of writing this paper in July 2010.

The second case is by NASA where it experienced the Challenger launch disaster followed by the Columbia disaster. Vaughan (1996, 2005) has analysed both failures and noticed a consistent institutionalised practice of underestimating failures as early warning signals with too much belief in the track record of past successful launches. (Smith, 2003) noticed that NASA concluded that both accidents were attributed to 'failures of foresight'.

The third case concerns Toyota. In January 2010, quality problem affected Toyota which led to a global recall of more than 8.5 million vehicles for various problems, including sticking gas pedals, braking software glitches and defective floor mats. And again on July 5th 2010, Toyota began recalling more than 90,000 luxury Lexus and Crown vehicles in Japan as part of a global recall over defective engines making it another setback for the automaker beset with quality problems (Kageyama, 2010).

The title of the paper has a funny first part which is 'Not just rearranging the deckchairs on the Titanic'. It implies that learning should not lead to doing something pointless or insignificant that will soon be overtaken by events, or that contributes nothing to the solution of a current problem. So why do failures happen? Are organizations really learning from failures? And in the context of a failure, how is learning realised? These are questions where there will be an attempt to answer them within the scope of this paper.

Learning from Failures as a Process Study

Process questions within the process study field are related to the evolution process of why and how things emerge, and develop, over time, which is different from variance questions focusing on co-variations among dependent and independent variables. An example of process studies at the organization level is (Balogun and Johnson 2004) research of how middle managers make sense of change as it evolves. Another example is (van de Ven and Poole 1995), where the authors focus on the temporal order and sequence in which selected managerial or organizational phenomena develop and change over time, which contributes to the process theories of organization and management. (Tsoukas and Chia 2002) suggests that processes of change are continuous and inherent to organizations. This is particular helpful when investigating how failures emerge, develop and the learning that can occur. In this paper there is an attempt to address a strand of process studies of organizational innovation and change that examines how individuals and organizations learn from failures, and how disasters unfold over time.

Within the field of organizational learning there is a view that incremental learning, tends to be control oriented which seeks to maintain predictable operations, minimize variation, and avoid surprises, whereas radical learning tends to seek to increase variation in order to explore opportunities and challenge status quo (Caroll et al., 2002). Failures are enablers for change in the status quo. They challenge the status quo and old assumptions. They force decision makers to reflect on what went wrong, why and how (Morris and More 2000).

The traditional school of thought tends to emphasize learning from successes (Sitkin, 1992), (McGrath, 1999), and (Kim, and Miner 2000). Whereas, a recent trend of research has begun to explore whether organizations can also learn from failures of other organizations, or from failures that occur within the same organization (Carroll *et al.*, 2002), (Haunschild, and Sullivan, 2002), (Denrell, 2003), (Haunschild and Rhee, 2004), (Kim, and Miner 2007), (Chuang *et al.*, 2007), (Desai, 2008), and (Madsen, and Desai 2010). Experience with failure has proven by Madsen and Desai (2010), and Haunschild and Rhee (2004) to be more likely than experience with success to produce conditions for experiential learning, the deriver to challenge existing knowledge, and the ability to extract meaningful knowledge form experience.

Rare events and organizational learning has traditionally been treated on the margin of the mainstream research on organizational learning, as they are often set aside as statistical outliers (Lampel *et al.*, 2009). However, a recent special issue in Organization Science in 2009 has been dedicated to the issue of rare events and organizational learning. In this work, Lampel *et al.* (2009) proposed a taxonomy of learning that is based on two categories; the impact of rare events on the organization, and their relevance for an organization. Based on those two factors, they identified four classes of events; transformative, reinterpretative, focusing, and transitory. They then mapped the works of Beck and Plowman (2009), (Christianson *et al.*, 2009), (Madsen, 2009), (Rerup, 2009), (Starbuck, 2009), and (Zollo, 2009) in to those classes. This paper is about disasters that have had high impact and hence they may be regarded as either transforming or focusing depending on their degree of relevance to a specific organization. However, there is also a provision of generic lessons that are applicable to wide range of industries.

It is generally accepted that learning from failures is a difficult process to assess. Few authors have attempted to define it. Organizational learning has been defined by Madsen and Desai (2010) as any modification of an organization's knowledge occurring as a result of its experience. But again it is acknowledged that a change in organizational knowledge is itself difficult to observe. Subsequently, there has been a trend in research which argues that changes in observable organizational performance reflect changes in organizational knowledge (see Baum and Dahlin 2007 and Argote 1999). Another line of research attempts to explore ways of learning. For example, (Carroll *et al.*, 2002) proposed a four-stage model of organizational learning reflecting different approaches to control and learning. Also, (Chuang *et al.*, 2007) developed a model of learning from failure of health care organizations that highlights factors that facilitate learning from failure.

Learning from disasters in the context of socio technical systems failures have been studied by the pioneering work of Turner (1978) on man-made disasters, Toft and Reynolds (1997) on identifying of a topology of isomorphism, and Shrivastava *et al.* (1988) on analysing symptoms of industrial crises, and a proposed a model of socio-technical crisis causation based on either 'HOT' events initiators of failures (Human, Organisational, and Technological) or 'RIP' events accelerators of failures (Regulatory, Infrastructural, and Political). This was then extended by the work of Smallman (1996) in challenging traditional risk management models, Pidgeon and O'Leary (2000) in investigating the interaction between technology and organizational failings, and Stead and Smallman (1999) in comparing business to industrial failures, where the issue of interaction between technology and organizations have been further analysed and applied to the banking sector.

Learning from failures has been researched as an outcome rather than the process itself. It has been argued by (Madsen and Desai 2010) that learning from failures is related to improved performance. It has also been argued by (Barach and Small 2000) that learning from near misses helps in redesigning and improvement in processes. Learning has been defined by Haunschild and Rhee (2004) to be, in the context of car recalls, as a reduction in subsequent voluntary recalls. Stead and Smallman (1999) proposed the process of a crisis as an event cycle consisting of a set of *pre-conditions*, a *trigger* event, the *crisis* event itself, a period of *recovery*, which then leads to the phase *learning*. In the proposed 'crisis cycle' theory they attempt to ground it based on the works of Turner (1997), Shrivastava (1998), Pearson and Mitroff (1993), and Pearson and Clair (1998), Case studies of accidents have been presented by Kletz (2001) in an effort to learn from accidents by carrying out a thorough analysis of the causes of the failures. Therefore, the majority of literature, apart from the few ones mentioned above, tends to focus on what is learning from failure as an outcome rather than why disasters occur. Also there is a lack of research work on how the learning process can emerge or how to use models that can facilitate the process of learning from failures and extract generic lessons. There is an argument that using a model to analyse investigate an accident may distract people who are carrying out the investigation as they attempt to fit the accident into the model and this may limit free-thinking (Kletz, 2001). We agree with this line of thinking and therefore argued in this present paper that the use of hybrid of models overcomes this issue as such an approach will lead to limiting the assumptions inherit in such models as they will cancel each other. This line of thinking will be expanded later on in the paper.

The above discussion shows that there are several layers that explain the current related theories. Therefore, it is proposed to use an onion of layers as in figure (1) that demonstrates where our paper fits with respect to those different theories. These circles are like layers of an onion, following (Saunders *et al.* 2003) approach of the research onion. So, to summarise, the proposed approach is oriented towards process oriented questions rather than variance questions, that focuses on radical learning rather than control learning, and that is based on failures rather than on successes.

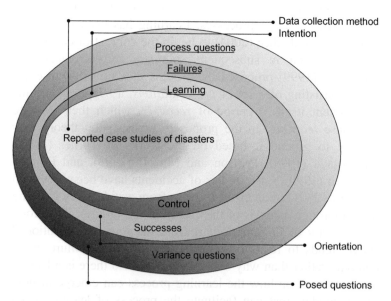

Figure 1: Research focus – Adapted from the Research Onion by (Saunders et al. 2003).

In order to address the issue of learning from failures we propose a framework model of learning from failures based on the analysis of four case studies related to reported disasters. We will apply techniques that facilitate learning from failure. We will then discuss why organizations need to learn from failures, and how, by identifying key generic lessons. Finally, we propose a framework of the learning process. Case studies chosen are related to: i) The Titanic Disaster, ii) The BP Texas City Incident, iii) The Chernobyl Disaster, and iv) NASA's Space Shuttle Columbia Accident.

Case Studies:

Choice of Case studies and research methodology

The case studies were chosen to demonstrate learning through the proposed framework of feedback to design, using advanced techniques, and interdisciplinary generic lessons. Case studies were also chosen from different industries and to some extent, are believed to have had a major effect on the whole industry where the case study has occurred. The reason for choosing reported case studies of major disasters is based on the arguments of (Greening and Gray, 1994) and (Desai, 2010) in that catastrophic failures are well observed and this visibility tends to encourage organizations and their members, as well as policy makers, to learn from those incidents. Although this methodology is limited by the reliance on investigation by others, it provides a fascinating body of research as outlined by (Wood *et al.*, 1994). The psychologist Lauren Slater argues that: *"We most fully integrate that which is told as a tale."* Therefore the analysis of these real-life case studies is important to demonstrate how subjectivity that relies on opinions of experts can be turned into a modelling approach that can ensure repeatability and consistency of results.

Accordingly, all case studies were constructed to follow a certain format in terms of first there is a description of what happened by presenting the sequence of the events that led to the disaster. This is followed by investigating the technical and the logic cause of the failure. Then the consequence of the failure is presented. Then the tools used for the analysis are presented. Finally, a set of generic lessons and recommendations are provided in order to prevent future system failure.

The two techniques that have been used in those case studies are fault tree analysis (FTA), and reliability block diagrams (RBD). Both techniques complement each other since the outcome of the FTA model is used as an input to produce the RBD model. Both techniques are described in Appendix A. We have analyzed the information provided in every case study and then developed a FTA and a RBD model for each case showing how the models were constructed using the information contained in the narrative of the case study.

The fault trees used here are more of a general logic tree than a strict classical fault tree which has traditionally been used to analyse failures at equipment level. The reason is that a disaster is depicted as a top event. This is normally not strictly traditional in the usual fault tree sense, i.e., a well-defined subset of a sample space, to which a well-defined binary indicator variable can be attached i.e. 1 for occurrence, 0 for non-occurrence. The advantage of using the proposed approach is that it offers richness to the model so that both subjective judgement and objective evaluation measures are taken into account and hence can act as a good mental model.

Although there is an issue that FT and RBD are complementary, since the outcome of a FT is used as an input to produce RBD, it is acknowledged that this may be questionable in some cases. Usually FT and RBD are alternative techniques, in that FT compute failure probabilities while RBD compute reliability which is the probability of not having failures. Moreover, considering that an AND gate in a FT is equivalent to a parallel arrangement of blocks in a RBD and an OR gate is equivalent to a series of blocks, explains why RBD and FT are more equivalent and alternative tools rather than complementary. Hence there is a view that they can be used to compute and express similar things and the proof is that RBD are drawn starting from the FT and mainly describe the same thing even if with a different picture. This may help to visualize things differently but should not lead to more confusion but rather a better comprehension of the problem. Our response to this argument, and choice to use both techniques, is based on the fact that RBD shows clearly areas of vulnerability when nodes are arranged in a series structure, whereas relatively safe areas are noticed when nodes are in parallel. This is not easily depicted from the FTA where it is difficult to arrive to those conclusions from a hierarchical structure of OR and AND gates. It can also be argued that both techniques are synergistic instead of redundant as one cannot draw a RBD unless a FTA is constructed first. Here we emphasise that RBD is different from a process layout which can be drawn directly, it is more about the root causes that led to the failure mode at the apex of the fault tree, and hence RBD is an outcome of FTA. Moreover, both techniques produce different outputs. For example, if data is available about each failure event one can then derive the total reliability of the system from the RBD, but this derivation of total system's reliability cannot be arrived at from the FTA. So to summarise, FTA provides the know-how about the logic cause of the failure (hence the use of logic gates), whereas RBD identifies areas of vulnerability and is capable of calculating total reliability given data about reliabilities of individual events.

We also acknowledge that there could be other techniques that can be used for learning from failures such as for example the Analytic Hierarchy Process (AHP) as demonstrated in analyzing the failure of the Concorde aircraft accident (Davidson, and Labib, 2003). However in this paper, we have applied the same tools for every case in order to demonstrate the usage of such tools across a variety of case studies and hence show their generic applicability to a wide range of cases. Specifically, through the case studies we will demonstrate the use of simple analytical tools which managers can find them useful to support the decision making process and the design of measures that can lead to an improve in the overall safety performance.

Case Study 1: Titanic the Unsinkable Ship That Sank

What Happened: The Titanic was considered the world's biggest and most luxurious ship at its time. The Ship hit the iceberg at such high speed that caused the resulting failure of the superstructure to be catastrophic. The iceberg caused a rip to the hull of the ship and damaged 5 of 16 watertight compartments.

Titanic sank after 2 hours and 40 minutes after the crash. On board were 2,228 between passengers and crew.

Technical and Logic of the Failure: It is believed that many factors have led to the disaster, but most notably they were attributed to wrong decisions made (*human factor*), *low visibility condition*, limited resources as what was available was insufficient for detection of iceberg (lack of technology and equipment), material weakness / metal failure (technology not available for *quality control and testing*), *inefficient evacuation procedure* and *insufficient life boats* (the life boats capacity was only 1,178 people), and too much belief in the unsinkable ship.

Consequences and Severity: Human loss of 1,300 people, loss of ship $7.5 m, reputation of White Star Line was badly damaged, and massive insurance premium increase.

Fault Tree Analysis and Reliability Block Diagram for the Titanic Disaster:

The FTA shown in figure (2) shows that there are three main factors that have caused the disaster to occur, namely; a) insufficient passengers evacuation, which includes insufficient life boats, not enough time to evacuate, and lack of emergency procedures; b) Iceberg collision due to poor visibility or high speed; and c) the collapsed structure due to either in appropriate material or manufacturing processes. Note that the OR logic gate denotes whether one factor was sufficient, and the AND logic gates signifies that all factors are needed to happen simultaneously.

The RBD in figure (3) is derived from the FTA where each AND gate is arranged in a parallel structure and each OR in a series structure and we normally start modelling from the top of the tree and cascade downwards. A good example of an AND gate is redundancy where all systems (main and back up) need to fail for the whole system to fail. So an AND gate is a more reliable configuration than an OR gate. In the RBD model of the Titanic, it is clear that the second parallel line of insufficient passengers evacuation is the most vulnerable as all boxes are arranged in series, whereas in the other lines there is some parallel structures. Here we assume that boxes carry

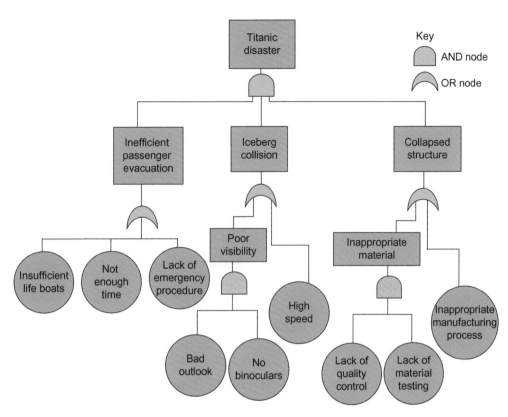

Figure 2: Fault Tree Analysis (FTA) of the Titanic Disaster.

equal weights since the actual vulnerability of a parallel line depends on the actual probability values of the blocks they include and not only on the way they are arranged. Although such probability values would enhance the calculation, it is often difficult to get this information in a disastrous situation.

Proposed Improvements and Generic lessons:

After carrying out the analysis using FTA and RBD one needs to reflect upon what could have been done differently to prevent the failure. A significant factor is human adherence to safety procedures which includes concurrence of decisions made and training the people about carrying out safety procedures especially with respect to passengers evacuation. Another important factor is the use of technology and equipment

Reliability Block Diagram (RBD) for the Titanic Disaster

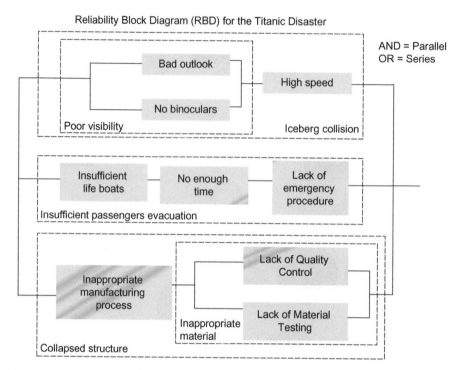

Figure 3: Reliability Block Diagram (RBD) of the Titanic Disaster.

available. This may include real time observation and safe running speed. Regarding poor design, there is a need for re-designing the ships to remain afloat after major impact and to review best available technologies. In addition, there is a need to re-design the ship superstructure using for example concurrent engineering methodology and employing multiple criteria evaluation for re-design.

Case Study 2: The BP Texas City Incident

What Happened: On March 23rd, 2005 disaster struck at BP's Texas City oil refinery when a series of explosions and fires resulted in the deaths of 15 people and injured more than 170. The explosions are believed to have started during the start-up of an isomerization (ISOM) process unit (BP, 2005). BP's Texas City refinery is BP's largest and most complex refinery with production levels up to 11 million gallons of gasoline per day and having a rated capacity of 460,000 barrels per day (bpd). This is in addition to the production of jet fuels and diesel fuels. The refinery's 1200 acre site is home to 30 different process units with around 1800 permanent BP staff. It is BP's largest plant, and the USA's third largest refinery, until 1999 the refinery was owned and operated by Amoco and still used Amoco's safety management systems which were employed before 1999 prior to the merger of Amoco and BP (BP, 2005). As mentioned earlier the incident initiated in the ISOM process unit and it involved the Raffinate Splitter, Blowdown Drum and Stack (BP, 2005). It is important to mention that there have been a number of previous events involving ISOM process units mainly involving hydrocarbons leaks, vapor releases and fires. What really draws attention is that two major incidents took place few weeks prior to the fatal event on March 23 (Khan, and Amyotte 2007). The first which occurred in February 2005 was an incident of hydrocarbons leaking to the sewer during de-inventory of the splitter, the second occurred in the same month of the fatal event and was a serious fire. Before going on to investigate the causes and the sequence of events which led to the incident, it is important to form an under-standing of the nature of the process which resulted in igniting the first event.

Technical and Logic of the Failure: A brief description of the processing equipment and the process itself is presented here to aid in understanding the incident. A key unit is the Raffinate splitter; this consists of a single fractionating column with 70 distillation trays and a feed surge drum, fire heated reboiler and overhead condenser (Mogford, 2005). The column processes up to 45,000 bpd of raffinate from aromatic recovery unit. The blowdown system consists of relief pipework headers, pumps and a blowdown drum and stack. The purpose of this system is to receive, quench and dispose of hot liquid and hydrocarbons vapors from the ISOM vents and pumpout systems during upsets or shutdowns (Khan, and Amyotte 2007). The unit works by heating the feed hydrocarbons (pentane and hexane) in order to convert them into isopentane and isohexane to boost octane rating of gasoline. The ISOM unit performs the conversion as the raffinate section prepares the hydrocarbon feed into the isomerization reactor (Holmstrom *et al.*, 2006).

Sequence of events and Incident

Everything began when the operators overfilled the distillation column. The liquid is supposed to remain at a relatively low level in the column; however the operators overfilled the column nearly to its top. As a result a mixture of liquid and gas flowed out of the gas line at the top of the column; the mixture passed through the emergency overflow piping and was discharged from a tall vent hundreds of feet away from the distillation column (Hopkins, 2008). The disaster could have been avoided should a continuously burning flame been present at the top of the vent which is what the case should have been ideally. In this manner any gases or mixtures erupting from the vent would have been burnt in the atmosphere and would have not formed a cloud and resulted in the explosion, however such a flare system was not available. Thus a vapour cloud formed with the potential of an immense explosion (Hopkins, 2008). Examining the events prior to the final accident; the raffinate splitter was shut down for a temporary outage on 21, February, 2005 and was later on steamed out to remove hydrocarbons on 26-28 February. On March 21 the splitter was brought back into service and was pressurised with nitrogen for tightness testing, on the next night cold feed was introduced to establish feed drum and column levels. As work started on March 21 after normal operation tasks, temperature and pressure levels in the splitter were seen to be higher than normal, so the operators opened an 8 inch vent valve and witnessed steam like vapours coming out, after doing so pressure and temperature dropped to normal levels (Khan, and Amyotte 2007). As flow of raffinate product commenced the shift supervisor suggested opening the 1.5 inch vent valve to vent off nitrogen, after shutting the vent again pressure peaked to 63 psig as a result the overhead relief valves opened to feed to into the unit which resulted in vapours and liquid coming out at the top of the stack (Khan, and Amyotte 2007). Hydrocarbon liquid and vapour were discharged to the blowdown drum which led to overfilling of the drum leading to a geyser like release from the stack. The final result was a series of multiple explosions and fires as the liquid hydrocarbons pooled on the ground and released flammable vapours ready to be ignited (Holmstrom et al., 2006). As mentioned earlier there was no flare system present, this is because the blowdown system was an out of date design installed in the 1950's and was never connected to a flare system to safely combust any process released vapours. It is worth mentioning that the accumulated vapour cloud is believed to have been ignited by a truck engine left idling in the area.

Investigation

According to BP's final investigation report (BP, 2005), there were four critical factors without which the incident would not have occurred or at least its consequences minimized, these four factors are:

Loss of containment: This involves the action taken or not taken which lead to overfilling of the raffinate splitter and eventual over pressurization and pressure relief. The end result was liquid hydrocarbon overflowing the stack which caused the vapour cloud. Elements which contributed to overflowing of the splitter and eventual pressure relief were the very *high liquid level* and *high base temperature* in addition to the *late start-up of the heavy raffinate rundown*. The high liquid level and normal tower pressure ended up surpassing the relief valve pressures. It is expected the stopping the feed, increasing the offtake or reducing the heat input during earlier stages or a combination of all three would have prevented the incident (Mogford, 2005).

Raffinate splitter start up procedures **and application of knowledge and skills:** Not following specified procedure resulted in loss of process control, appropriate knowledge was not applied and there was no supervision during start up. Deviation from operating procedures resulted in overfilling of raffinate splitter; there was a 3 hour delay in starting the heavy raffinate splitter in addition to adding the heat at a higher rate than specified and before initiating heavy raffinate rundown. The maximum temperature used also exceeded that specified by the procedure. Any actions that were taken to resolve the situation were either late or inadequate and mostly even made the situation worse. Feed was never cut and no supervision was present pre or during start up. If procedure was followed, the incident would not have happened although there was a clear fault in the chain command and an ambiguity in roles and responsibilities.

Control of work and trailer sitting: During start up, personnel working elsewhere were in close proximity to the hazard. People were not warned or evacuated as they were assembled in and around temporary trailers. The blowdown drum and stack was not considered as a potential hazardous source in any of the site studies, hence trailers were located too close to the source. The people were not notified of the start up process or when the discharge occurred. The evacuation process was ineffective as alarms did not sound early enough. The location of trailers and poor communication increased the severity of the incident (Mogford, 2005).

Design and engineering of blowdown drum and stack: The use of blowdown drum and stack as a relief and venting system for the splitter near uncontrolled areas considering the several design and operational changes over time. The splitter was not tied to a relief flare system although there were a number of chances to do so; several changes were introduced to blowdown system which reduced its effectiveness. The use of a flare system would have reduced the impact of the accident (Mogford, 2005).

The causes to these critical factors vary from immediate causes to complex managerial and culture issues, these shall not be discussed here and focus will remain on the direct factors discussed already. In order to examine how these factors added up together, two different reliability and fault finding techniques will be discussed briefly next.

Fault Tree Analysis and Reliability Block Diagram for the BP Texas City Disaster:

The FTA of the disaster is shown in figure (4) where it is shown that the four main factors that have led to the disaster are: a) loss of containment, b) Start-up Procedure of raffinate, (c) Planning and control of work, in terms of Weak risk assessment, and d) Design and Engineering of blowdown drum and stack. All these factors have contributed to the accident and hence are linked with an AND gate at the top of the fault tree in figure 4. From the narrative of the case study it is clear that the loss of

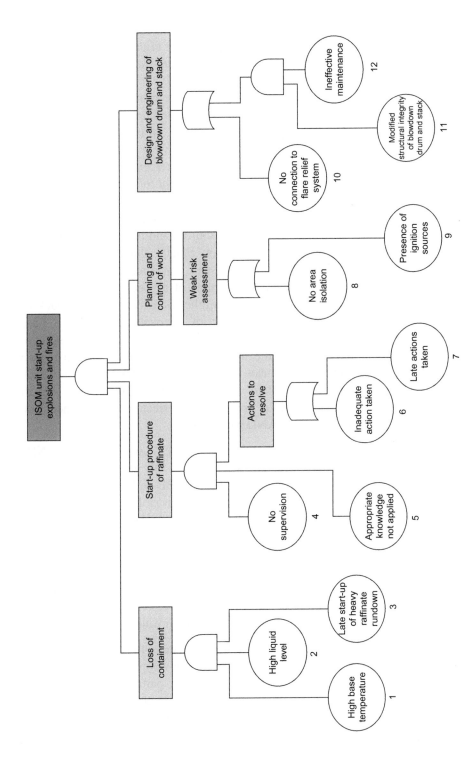

Figure 4: BP Incident Fault tree analysis.

containment is attributed to very *high liquid level* and *high base temperature* in addition to the *late start-up of the heavy raffinate rundown and hence all three factors are linked with an AND gate. According to the narrative of the case study, the* Start-up procedure of raffinate *is attributed to lack of* supervision during start up, appropriate knowledge was not applied, and actions taken to resolve the situation were either late or inadequate, and hence the actions are further extended using an OR gate. *Regarding planning and control of work this was primarily attributed to weak risk assessment which in turn was caused by either a weak* evacuation process in the form of area isolation or the presence of ignition sources and hence an OR gate is used to model those two factors. Finally the design and engineering of blowdown drum and stack was a major factor that contributed to the accident. An example of bad design was the fact that the splitter was not tied to a relief flare system. or the lack of structural integrity and lack of adequate maintenance procedures and hence an OR gate is used. The use of a flare system would have reduced the impact of the accident although there were a number of chances to do so. Several changes were introduced to blowdown system which reduced its effectiveness and this was coupled with the presence of ineffective maintenance procedures and hence an AND gate is used.

Now starting from the top of the FTA we model each AND gate as a parallel structure whereas each OR gate is modelled in a series structure. It is clear from the RBD in figure 5 that all four main factors that have led to the disaster happened at the same time and hence the parallel structure (AND gate). It might be a mere co-incident or an indication of many things that went wrong, a situation that can be explained as 'an accident waiting to happen'. It is also clear from the RBD model in figure (5) that items 8 (*no area isolation*), and 9 (*presence of ignition sources*) - where both are attributed to the existence of *weak risk assessment* - are structured in a series configuration which shows their vulnerability.

In order to prevent future system failures there is a need to address design integrity and develop better operating practices and procedures in the plant. In particular, the leadership team needs to establish clear common goals with rules and procedures, emphasizing teamwork and a one-site mentality and the leadership team must hold superintendents and supervisory employees

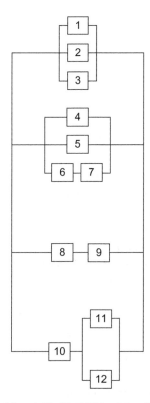

Figure 5: BP Incident Reliability block diagram.

accountable for implementation of rules and procedures. There is also a need to establish a risk awareness training program for all and to make sure that leadership responsibilities include facilitating hazard identification. There is a need to conduct a fundamental review of control of work process and practices.

Case Study 3: The Chernobyl Disaster

What Happened: In 1986, two explosions occurred at the newest of the four operating nuclear reactors at the Chernobyl site in the former USSR at an estimated loss of £ 3 billion. The full extent of the long-term damage has yet to be determined.

Technical and Logic of the Failure: The Unit 4 reactor was to be shutdown due to maintenance scheduled for the nuclear plant on 25 April 1986, and then authorities from Moscow decided to take this period as an advantage for carrying out an experiment within the plant. The staffs on duty were ordered to carry out the experiment in order to assess whether, in the event of a loss of power in the station, the slowing turbine could provide enough electrical power to operate the main core cooling water circulating pumps, until the diesel emergency power supply became operative. The basic aim of this test was to ascertain the cooling of the Core would continue even in the event of a power failure. This is ironic as the experiment was supposed to test the same failure mode that happened during the execution of the experiment itself. *"Unfortunately, there was no proper exchange of information and co-ordination between the staffs in charge of the test and the safety personnel in charge of the reactors, as the test would also involve the non-nuclear part of the power plant. Hence, there was no adequate safety precautions included in their test programme as the operating staffs were left un-alert to the nuclear safety implications of their electrical testing and its potential danger"*. (World Nuclear Association, 2009).

The test programme would require shutting off the reactor's emergency core cooling system (ECCS), as this component was responsible for providing water to cool the core in the case of an emergency, although subsequent testings were not greatly affected by this, but the exclusion of such component in performing such practice showed the level of ignorance the operating staffs had towards the safety of a very high class plant.

Thus, as the planned shutdown was getting in place, the reactor was then operating at about half power just when the electrical load dispatcher refused to allow further shutdown, as the power was needed for the grid. In accordance with the scheduled test programme, just about an hour later the reactor was then operating at half power even as the ECCS was switched off. Though not until about 23:00 on 25 April that the grid controller had to make a further reduction in power.

Causes of the Incidence:

The causes of this huge catastrophic disaster was actually investigated by the IAEA (International Atomic Energy Agency), which had to set up an investigation team known as the International Nuclear Safety Advisory Group (INSAG), which in its report of 1986, INSAG-1, gave the following as the causes of the accident;

- There were gross violations of operating rules and regulations; as within the course of preparation and testing of the turbine generator under run-down conditions using the auxiliary load, the operating staffs disconnected a series of technical protection systems and breached the most important operational safety condition for conducting a technical exercise.
- However, the aforementioned also gave rise to during the time of the accident many of the key safety systems were shut off, most notably the Emergency Core Cooling System (ECCS).
- Also the reactor operators disabled safety systems down to the generator which the whole testing programme was all about.
- Their main process computer, SKALA, was running in such a way that the main control computer could not shut down the reactor or even produce power.
- All control was transferred from the process computer to the human operators (IAEA, 1991, 2005).

Fault Tree Analysis and Reliability Block Diagram for the Chernobyl Disaster:

The FTA analysis shown in figure (6) shows that the two main causes of the disaster were attributed to errors by the operating staff and designers. According to the narrative of the case study, factors contributed to errors by operating staff are due to lack of training, lack of communication between safety and operating staffs, and violation of operating regulations; all linked with an AND gate as they occurred simultaneously. The lack of communication issue is evidenced by disconnection of main control from the reactor, the transfer from process to human operators, and the disconnection of technical protection system without the involvement of the operators. As for the issue of bad design, there are possible causes of the accident attributed to wrong void coefficient value, operating reactivity margin, main circulation pumps, or faulty sub-cooling system; hence joined by an OR gate.

The RBD model in figure (7) shows that in terms of the operating staff there was particular problem with lack of communication between safety and operating staff, as the more boxes added in series the less reliable the system becomes. The same applies

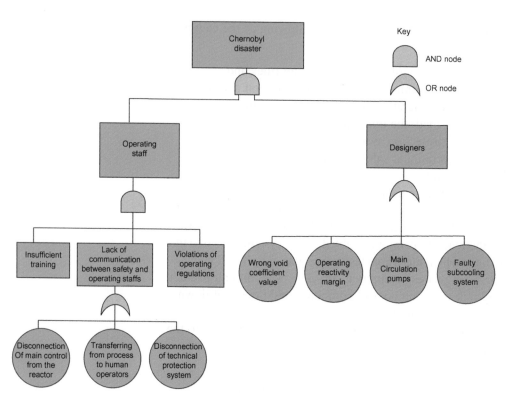

Figure 6: The FTA of the Chernobyl Disaster.

to the issue of bad design which is evidenced by four boxes in figure 7 connected in series.

Proposed Improvements and Generic lessons:

- Designers have to simulate their designs and carry out exhaustive testing.
- Operating staffs are required to be adequately and intensively trained as regards to safety, operations and troubleshooting.
- Communication between operations and safety personnel were very important to such operations.
- Consultation between designers, original equipment manufacturer (OEM) and the operations should be held in the case of testing.
- Operating staffs should be required to follow the design procedures.

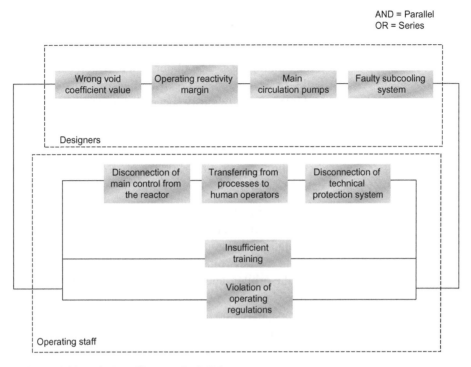

Figure 7: The RBD of the Chernobyl Disaster.

Case Study 4: NASA's Space Shuttle Columbia Accident

What Happened: The loss of Columbia was caused by damage sustained during launch when a piece of foam insulation which was the size of small briefcase broke off the main propellant tank under the aerodynamic forces of launch. The debris struck the leading edge of the left wing of the number 8 reinforced-carbon-carbon (RCC) tile, damaging the shuttle's thermal protection system (TPS).

The National Aeronautics and Space Administration (NASA) launched the space shuttle *Columbia* on its STS-107 mission on January 16, 2003. On February 1, 2003, as it descended to Earth after completing a 16-day scientific research mission, *Columbia* broke apart over northeastern Texas. The disaster occurred on 1st Feb 2003 at an altitude of 203,000 ft in the area above north central Texas. The shuttle was moving at a speed of 12,500 miles/hr. Solar wind was above 800 kms /sec (in the red zone). But average solar wind speed should be around 300-400 Kms/sec.

The communication was lost at 9.00 am EST. At 39 miles high, data was lost from the temperature sensors on the left wing followed by the data lost from the tire pressure sensors on the left main landing gear. The obiter in left wing began to break up and tanks of Hydrogen, Oxygen and propellants began to explode. Columbia has previously succeeded in its 27 missions to space.

Technical and Logic Cause of the Failure:

According to (Smith, 2003), NASA identified three broad categories of causes to the accident.

- Physical cause

  The physical cause was damage to *Columbia*'s left wing by a 1.7 pound piece of insulating foam that detached from the left "bipod ramp" that connects the External Tank to the orbiter, and struck the orbiter's left wing 81.9 seconds after launch. As Columbia descended from space into the atmosphere, the heat produced by air molecules colliding with the Orbiter typically caused wing leading-edge temperatures to rise steadily, reaching an estimated 2,500 degrees Fahrenheit (1400°C) during the next six minutes. The foam strike created a hole in a Reinforced Carbon-Carbon (RCC) panel on the leading edge of the wing, allowing superheated air to enter the wing during reentry. The extreme heat caused the wing to fail structurally, creating aerodynamic forces that led to the disintegration of the orbiter. So while landing, due to the increase in the temperature the obiter in the left wing broke down.

- Organizational causes

  The accident was rooted in the Space Shuttle Program's history and culture, including the original compromises that were required to gain approval for the Shuttle, subsequent years of resource constraints, fluctuating priorities, schedule pressures, mischaracterization of the Shuttle as operational rather than developmental, and lack of an agreed national vision for human space flight. Also, the crew people were not informed about this damage.

- "Broken safety culture" at NASA

  Schedule pressure related to construction of the International Space Station, budget constraints, and workforce reductions also were factors. The shuttle program has operated in a challenging and often turbulent environment. It is to the credit of Space Shuttle managers and the Shuttle workforce that the vehicle was able to achieve its program objectives for as long as it did."

Consequences and Severity: This mission claimed 7 crucial lives of elite NASA astronauts. Following the loss of Columbia, the space shuttle program was suspended. A huge financial loss was estimated of around $140 M USD. A Helicopter was crashed while in the search of debris claiming 5 lives. There was a huge property loss resulted due to falling of Debris and reimbursement of $50,000 was claimed.

Proposed Improvements and Generic lessons: NASA created the *Columbia* Accident Investigation Board (CAIB) to investigate the accident. They recommended to devise a plan to inspect the condition of all reinforced carbon-carbon (RCC) systems, since the board found current inspection techniques inadequate. RCC panels are installed on parts of the shuttle, including the wing leading edges and nose cap, to protect against the excessive temperatures of re-entry. They also recommended to take images of each shuttle while in orbit standard procedure. As well as upgrade the imaging system to provide three angles of the shuttle from lift-off to at least solid rocket booster separation. "The existing camera sites suffer from a variety of readiness, obsolescence and urban encroachment problems". So the CAIB offered this suggestion since NASA had no images of the Columbia shuttle clear enough to determine the extent of the damage to the wing. They also recommended conducting inspections of the thermal protection system, including tiles and reinforced carbon-carbon panels, and develop action plans for repairing the system. The CAIB report included 29 recommendations,15 of which the Board specified must be completed before the shuttle returns to flight status and also made 27 "observations" (CAIB, 2005).

Fault Tree Analysis and Reliability Block Diagram for the Columbia Disaster:

The FTA diagram in figure (8) and the RBD diagram in figure (9) show that in order to avert such a disaster, it is important to

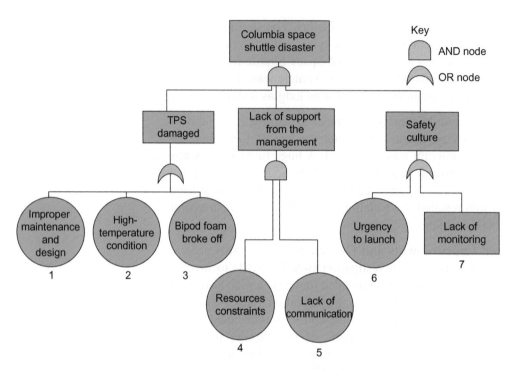

Figure 8: Fault Tree Analysis of the Columbia Space Shuttle Disaster.

improve the reliability of the system (as evidenced by the damage of the TPS), support from management, and safety culture. All three factors have contributed simultaneously to the disaster and hence have been linked using an AND gate in the FTA and accordingly as a parallel structure in the RBD model.

Proposed Framework Model:

We propose a framework model based on the outcome of those case studies of high profile disasters that learning can be addressed in three perspectives which are: i) feedback from the users (maintenance) to design, ii) the incorporation of advanced tools in innovative applications, and iii) the fostering of interdisciplinary approaches and generic lessons.

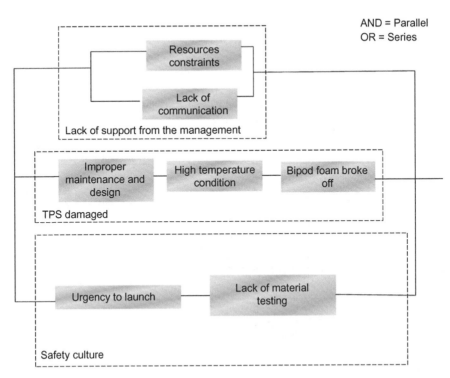

Figure 9: Reliability Block Diagram of the Columbia Space Shuttle Disaster.

The feedback from users to design is important as it provides lessons that are directly attributed to the specific case in question through analysis of possible failures paths and hence can help to improve reliability of the system through incorporation of redundant systems, improved procedures, training, increased factors of safety and so on. Generic lessons are useful for other organizations and industries to learn from. Generic lessons tend to come from teams with more average years of plant experience (Carroll *et al.*, 2002), and those years of experience tend to be accumulated in a tacit type of knowledge (Nonaka, 1994). The incorporation of advanced tools and techniques such as FTA and RBD are the facilitators that can help to identify and analyse root causes and derive lessons from failures; a process of transforming tacit to explicit knowledge.

Table 1: Proposed Framework Model of Learning Process as an Outcome of the Case Study Disasters

	Case 1: Titanic	Case 2: BP Texas City	Case 3: Chernobyl	Case 4: Columbia
Feedback from users (maintenance) to design.	Number of life boats and evacuation and emergency procedures.	Planning and control of work through proposer risk assessment.	Stronger communication between designers, engineers and operational personnel.	Safety aspects of on flight monitoring and repair.
Incorporation of advanced tools in innovative applications, such asFTAandRBD	a) FromFTA: insufficient passengers evacuation, which includes insufficient life boats, not enough-time to evacuate, and lack of emergency procedures; b) Iceberg collision due to poor visibility or high speed: and c) the collapsed structure due to either in appropriate material or manufacturing processes. b) From RBD: insufficient passengers evacuation is the most vulnerable.	a) FromFTA: the four main factors that have led to the disaster are: a) loss of containment, b) Start-up Procedure of raffinate. (c) Planning and control of work, in terms of Weak risk assessment, and d) Design and Engineering of blowdown drum and stack. b) From RED:items 8 (No area isolation), and 9 (Presence of ignition sources): where both are attributed to the existence of weak risk assessment, are structures in a series configuration which shows their vulnerability.	a) FromFTA: The two main causes of the disaster were attributed to errors by the operating staff and designers. b) From RBD: In terms of the operating staff there was particular problem with lack of communication between safety and operating staff.	a) FromFTA: It is important to improve the reliability of the system, support from management, and safety culture. b) From RBD: Importance of simulation to avert physical damage of the TPS system as it has a relatively high number of boxes in a series structure.
Generic Lessons and fostering of interdisciplinary approaches.	Safety procedures. Use of technology. Redesign of superstructures	Leadership. Risk awareness. Control of Work. Skill turnover. Workplace conditions. Rotation of management.	Design simulation. Adequate training. Improved communication. Consultation between stakeholders. Adherence to procedures.	Decentralisation and independence of safety budget. Recertification prior to operation. Provision of adequate resources for long-term programs to upgrade design and reliability of systems.

Overall Generic Lessons:

In this section we derive generic lessons that can help organizations, and managers, to understand reasons for failures. We use the term generic lessons to imply lessons that can be used by other sectors, as well as lessons that can be used by governments (Myddelton, 2007) and policy makers at a strategic level.

The lessons derived from the detailed empirical analysis of previous case studies can be argued to be quite specific to those particular accidents and one can argue that it is difficult to replicate the exact environment where those accidents have occurred and hence there is limited scope of learning any lessons as it is unlikely that similar accidents will occur. We respond to this argument by providing some generic lessons of how such disasters can help us to understand why failures happen and how to prevent them. The tools presented can help to avoid such disasters and reduce their consequences.

The area of reliability and maintenance (R&M) engineering deals with failures, and hence it is can be extended to learning from failures in the context of disasters and vice versa. It is thus useful to compare disasters to traditional R&M disciplines, and then suggest possible ways of dealing with disasters, based on lessons learnt from the advances in R&M. Conversely, generic lessons learned from disastrous failures can help R&M concepts advance even further (Labib, 1999).

Generic Lesson 1 – Too Much Belief in Previous Successes:

We argue that experience with success can be, and has been, counterproductive. For example, too much belief in the unsinkable ship made the Titanic disaster in 1912 to come as a major surprise. The Titanic was perceived in 1912 'as the safest ship ever built' (The Sinking of the Titanic, 1912, 2000). This false perception led to the fatal error of providing insufficient life boats. The life boats capacity was for only 1,178 people which is about half of her 2,200 passengers and crew on board. In fact the life boats were not intended for those who were on board but rather to rescue survivors of other sinking ships because of the too much belief in the 'unsinkable' ship and that the Titanic was herself considered as a life boat and hence there was no need to install lifeboats, which took up valuable deck space. More recently, much belief in the success of previous shuttle missions caused NASA to ignore warning signals related to both the o-rings damage prior to the Challenger disaster in 1996 due to cold weather before launch, and again on the fuel tank foam losses prior to the Columbia disaster in 2005. According to the investigation report *'NASA's safety culture had become reactive, complacent and dominated by unjustified optimism'*.

Generic Lesson 2 – Coping with Growth:

Indeed the ability, or inability, to cope with a high rate of growth may be another factor that can contribute to disastrous failures. Toyota has been regarded as the company against which other organizations benchmark their standards. This is evident from the best selling management book, The Machine that Changed the World by (Womack, Jones et al. 1990), which is based on Toyota Production System (TPS). Nevertheless, the recent failures that have led to the massive recall of Toyota vehicles in the US, Europe, and China, in 2010, is attributed to the inability to cope with the rate of growth of the company, which over-took GM in April 2007 to become the World leader in market share. Could BP have come across same circumstances of rapid growth given the soaring of its share prices prior to the accident in 2010? One always expects that catastrophic failures are inversely proportional to resources available. So in good times, resources are more available and hence more investment goes into safety measures. However, it appears that the relationship between company's prosperity and hazard is probably rather of a U-shape one, where the two extremes of bad or good times are not good news as far as safety is concerned. It is difficult to assess this hypothesis but it seems that history teaches us that in bad times resilience, and even faith, tend to be stronger than in good times.

Generic Lesson 3 – Misconception of Fashionable Paradigms:

The misconception of fashionable paradigms might be a reason of why man-made disasters happen. For example, through the first author's experience with organizations, it was observed that the concept of Lean Thinking, or Lean Manufacturing, has often been misinterpreted as a cost cutting exercise; 'taking the fat out of the meat'. Organizations, in an attempt to identify and minimise waste (Muda in Japanese Language), might cross the line and end up unknowingly sacrificing safety. Cutting cost is just one aspect of lean thinking which attempts to find and eliminate muda. But also lean is about adding value and stream lining a process. Could any of the above-mentioned organizations sacrificed safety in the name of lean? Mis-conception of paradigms has always been a dangerous affair. We have seen this before when Business Process Engineering (BPR) was a fashion, and it ended up by people mocking the acronym of BPR as 'Bastards Planning Redundancies'. Next generation manufacturing paradigms have been investigated by (Alvi, and Labib, 2001). For a good analysis of what is a fad and what makes a paradigm see (Towill, 1999).

Generic Lesson 4 – Legislations:

The concept of Corporate Social responsibility (CSR) emphasis the social aspect but perhaps the "S" in the acronym should be referred to as "Safety" as well rather than embedded among the social responsibilities. (Chouhan 2005) made a comparison between two sister plants owned by the same company; Union Carbide, and argues that the significant differences between the West Virginia, USA plant and the Bhopal, India plant show the callous disregard of the corporation for the people of the developing countries. It is just this year, 2010, that officials responsible for Bhopal were prosecuted for an accident that occurred back in 1984. There seems to be a need for something similar to the three strikes and out law at a corporate level in terms of serious breaches of safety. The Corporate Manslaughter and Corporate Homicide Act 2007 in the UK is a good step in this direction, where for the first time, organizations can be found guilty of corporate manslaughter as a result of serious management failures resulting in a gross breach of a duty of care.

Generic Lesson 5 – The "I operate, You fix" attitude:

In old-fashioned maintenance, a prevailing concept among operators is "I operate, You fix". In other words, maintenance is the responsibility of the maintenance department and operators should deal only with the operation of their own machines. Contrasting this attitude with the situation when dealing with a disastrous situation, one can easily notice that the majority of people feel that the responsibility of dealing with a disaster lies only with the health and safety or quality department. It is important for everybody, especially top management, to be aware that a disaster is not just another accident issue and that their direct involvement is necessary (Labib, 1999).

Generic Lesson 6 – No News is Good news:

In old-fashioned maintenance, the prevailing attitude is to make minimal effort for prevention, and the main mode of work is corrective maintenance. One famous saying is "If it ain't broken, don't fix it". This implies a passive attitude towards performing any prevention activity, a wait-and-see situation. In most organizations, a preventive maintenance schedule could be well-prepared but usually not implemented. The reason is that breakdowns take priority over preventive actions. In the context of disasters, it seems that the same attitude prevails, especially in the private sector, as organizations are more worried about current rates of exchange, interest rates and current market shares than about the prevention of a disaster (Labib, 1999).

Generic Lesson 7 – Bad News Bad Person:

In old-fashioned organizations, the manager always prefers to hear good news. Someone who brings bad news about the malfunction, or even the expectation of it, is considered an under-performer. It is similar to the old days in the army, when a soldier bringing news about casualties would be in danger of being shot, as if it were his own fault. In the context of a disaster, it is common to find the situation that CEO will ask, say his head of IT Department, about the potential problems in a threatening mode, and will receive the answer within hours that everything is under control.

Generic Lesson 8 – Everyone's Own Machine is the Highest Priority to Him:

In traditional maintenance, every machine is the highest priority to its operator, and the one who shouts loudest gets his job done. In old-fashioned maintenance, it is a crisis mismanagement situation where priorities are ranked into three options: urgent, red hot, and do it or else. This lack of a systematic and consistent approach to setting priorities tends to be an important feature when dealing with a disaster. Setting priorities should attract the highest priority among different approaches to dealing with any potential disasters. Questions such as who sets priorities, what criteria are considered, and how to allocate resources based on prioritisation, urgently need to be addressed. The decision making grid (DMG) is a method used to prioritise machines based on the combination of their severity (measured by downtime) and occurrence (measured by frequency of breakdowns), provides various maintenance strategies in response to the relative sate of the machines. See (Labib, 2008), (Labib, 2004), and (Fernandez *et al.*, 2003).

Generic Lesson 9 – Solving a Crisis is a Forgotten Experience:

In old-fashioned maintenance, it is often the case that solving a breakdown problem does not get recorded or documented. The reason is that it is often considered a bad experience that is usually forgotten. An analogy here is to imagine that you ask several applicants for a job to write their CV. It is most likely that they will all write about their achievements and none of the applicants will attempt to write about their bad experiences. Nobody would be proud to mention them. On the other hand, modern maintenance techniques stress that a crisis is an opportunity for investigation, and failures should be well documented for future analysis. Unfortunately, in many near misses situations, organizations, and people, do not reveal their experiences with potential failing equipment or mistakes. One reason for that might be the fear of losing lawsuits and insurance claims. A good example is in the healthcare system (Barach and Small, 2000). It would, however, be beneficial to both organizations and individuals to be able to easily access databases of mistakes or near misses (Kim and Miner, 2007).

Generic Lesson 10 – Skill Levels Dilemma:

In the maintenance function, the designer of the machine is not usually the one who fixes it, and surprisingly, might not even have the ability to do so. Skills needed to restore particular equipment include functions such as diagnostics, logic fault finding, disassembly, repair and assembly. Depending on the level of complexity of particular equipment, as well as on the level of complexity of the function that needs to be carried out, the necessary skill level can be determined. In a crisis, the issue of skill levels needed is a major dilemma. The reason for that is that a disaster is a multi-disciplinary problem as it can span various fields such as information systems, maintenance, decision making, and crisis and risk management. Hence it needs a synchronised, multidisciplinary team approach (Labib, 1999).

Attributes of the Generic Lessons:

The above-mentioned lessons can be attributed to three main issues; a) responsibility, b) communication, and c) priority. There are at least five lessons that are attributed to each issue. In other words, as shown in Table 2, each column has at least five ticks. However, each generic lesson mentioned affects one or more of the three categories (columns). In this categorization, we assume that priority is defined in the broader sense, that is, it covers areas of criticality and resource allocation. There are at least five generic lessons that are attributed to each issue. In other words, as shown in Table 2, each column has at least five ticks. However, every lesson mentioned affects one or more of the three categories (columns). Take for example one of those categories, say *responsibility*. One can find that, in terms of responsibility: *legislations, the "I operate, you fix" attitude, bad news bad person, solving a crisis is a forgotten experience, and the skill levels dilemma*, are all issues that have been addressed in R&M. All those lessons related to responsibility share a common theme: trying to answer the main question about who is responsible for doing what. One could similarly view lessons learnt in terms of communication

Table 2: Generic Lessons Learnt from Failures

Generic Lessons:	Responsibility	Communication	Priority
Generic Lesson 1 – Too Much Belief in Previous Successes.		√	√
Generic Lesson 2 – Coping with Growth.			√
Generic Lesson 3 – Misconception of Fashionable Paradigms.		√	
Generic Lesson 4 – Legislations.	√		√
Generic Lesson 5 – The "I operate, You fix" attitude.	√		
Generic Lesson 6 – No News is Good news.			√
Generic Lesson 7 – Bad News Bad Person.	√	√	
Generic Lesson 8 – Everyone's Own Machine is the Highest Priority to Him.			√
Generic Lesson 9 – Solving a Crisis is a Forgotten Experience.	√	√	
Generic Lesson 10 – Skill Levels Dilemma.	√	√	

and prioritisation. Communication covers links between employees of different functions in the organization as well as within the same function. It is also about links across the supply chain, between companies and their suppliers, and between companies and their customers. With regard to a disastrous situation, if any of those links is weakened, then possibly one, or a combination of the five features related to communication (Table 2) would occur. Prioritisation can be addressed from two views: a) mode of work; and b) degree of criticality. One view is the priority of the mode of work, for example preventive versus corrective modes as shown in the generic lesson *no new is good news*, or allocating resources in response to changes in the market, as shown in generic lesson *coping with growth*. Another view of priorities is that it involves a degree of criticality of equipment as shown in generic lessons *too*

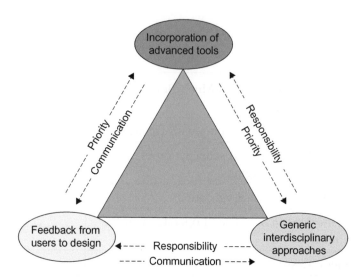

Figure 10: The unified learning from failures process model.

much belief in previous successes and *everyone's own machine is the highest priority to him.*

We propose integration of both models; the proposed model of the learning process as an outcome of the case study disasters as shown in table 1, and the generic lessons learnt from failures, as illustrated in table 2, into a unified framework model as shown in figure (10) below.

Concluding Remarks

In this paper a proposed framework model has been outlined in Tables 1 and 2 and illustrated in Figure 10 as an overall learning model from disasters. The novelty in this paper is four-fold. First, through the proposed framework model that is based on the outcome of case studies of high profile disasters. This model shows that learning can be addressed in three perspectives which are: i) feedback from the users (maintenance) to design, ii) the incorporation of advanced tools in innovative applications, and iii) the fostering of interdisciplinary approaches and generic lessons. The second aspect of novelty is the generic lessons identified that can help organizations, and managers, to under-stand reasons for failures, where ten lessons have been identified and summarised in Table 2. The third aspect of novelty is identi-fication of three attributes that affect all generic lessons, namely; a) Responsibility, b) Communication, and c) Priority. In Table 2 an attempt has been made to map attributes to the generic lessons. The forth aspect of novelty is the proposed integration of both models; the proposed model of the learning process as an outcome of the case study disasters as shown in table 1, and the generic lessons learnt from failures, as illustrated in table 2, into a unified framework model as was shown in figure (10) which combines aspects of direct feedback from the specific disaster to the designers of the system, the incorporation of advanced tools such FTA and RBD, and the ability to extract generic lessons that are of benefit to other disciplines or indus-tries. It also uses the aspects of responsibility, communication and prioritisation as the main derivers to facilitate the learning process.

The overall message of this paper is that it provides a frame-work for a holistic view regarding coping with failures. It is acknowledged that individual human beings are naturally pro-grammed to learn, whereas organizations are not (Carroll *et al.*, 2002). Therefore our approach provides a framework by which both individuals and organisations are able to learn from failures. Moreover, it provides some strategic issues to policy makers such as the importance of balancing the punishment and incentive systems and that they should also be proportional to the significance of hazardous incidents.

Specifically, through the case studies we demonstrated the use of simple analytical tools which managers can find them useful to support the decision making process and the design of mea-sures that can lead to an improve in the overall safety perfor-mance. However, it is important to note that failures in general and disasters in particular can stimulate a blame culture that can act as a barrier to learning from failures.

Through the construction of the fault tree, and block diagrams, of the events we were able to model each of the disasters by focusing on each incident as the undesired event to study at the top of the tree. This has helped to obtain an understanding of the system and to systematically evaluate the root cause of the events by cascading down into further levels of the tree. Through this analysis one is able to control the hazards identified. However, experience gained from applying the same technique to a variety of disasters has identified some limitations such the inability to analyse the complex order of components failures with respect to time. Also the hierarchical construction in one direction has an implicit assumption of independence among different aspects of the tree at the same level. So interdependencies among components within a system were not captured. Despite those limitations we were able to gain a better understanding of the process of failures as it provides a useful graphical representation of hierarchical analysis of failure modes, since it provides a map, and a mental model, that can help to understand the logic of the failure concerned. Also, it can be used for diagnostics and fault finding. This is because it follows a logical and systematic process of breaking down a complex problem into its root causes. Moreover, it allows quantitative analysis to be performed i.e. the probability of system failure, frequency of failure, system downtime and expected number of failures in x years. It helps to focus attention on the top event, a certain critical failure mode or an event, and then cascades deductively. It helps to identify the items of the system which are important the particular failure being studied and it provides ways of analysing human, software, procedural, and other factors in addition to the traditional physical parts of the system. Hence it provides opportunities to deepen our understanding and produce organizational change and learning.

Appendix A: Fault Tree Analysis (FTA):

A fault tree is a logical diagram which shows the relation between system failure, i.e. a specific undesirable event in the system as well as failures of the components of the system. The undesirable event constitutes the top event of the tree and the different component failures constitute the basic event of the tree. In other words, FTA is a top-down approach to failure analysis, starting with a potential undesirable event (accident) called a TOP event, and then determining all the ways it can happen. The causes of the TOP event are "connected" through logic gates; we only consider AND-gates and OR-gates. Although there are other types of gates their use in modeling reliability diagrams tends to be limited and hence not considered in our analysis as the majority of problems can be modeled by either AND or OR gates.

Basic events are those associated with human errors, equipment failure and environment interference. FTA provides a logical representation of the relation between the top event and those basic events. From a design perspective, this technique could give indications of how a system could fail which is of equal importance to how a system will operate successfully.

In order to achieve a rational analysis a number of steps should be followed in constructing the fault tree (Xing and Amari, 2008). First of all, the undesired event to be analysed should be defined clearly. The boundary conditions for analysis should be defined as well; this will define what parts are included in the study, the environmental influences included and the level of resolution and depth of the analysis. For an OR gate, occurrence of one or more of the lower events will result in the next level event, for an AND gate, all lower events must occur for the upper level event to take place (Krasich 2000). Symbols for OR and AND gates are shown in table 1.

Table 1 OR and AND logic gates (Krasich 2000)

Symbol	Symbol name	Description	Reliability Model	Inputs
	OR gate	The output event occurs if any of its input event occur	Failure occurs if any of the parts of the system fails-series system	≥ 2
	AND gate	The output event occurs if all input event occur	Parallel redundancy, one out of n equal or different branches	≥ 2

For AND and OR gates the probabilities are calculated as follows:

$$P_{(AND\_Gate)} = \prod P_{(input\_i)}$$

$$P_{(OR\_Gate)} = 1 - \prod (1 - P_{(input\_i)})$$

For AND Gate, the following logic applies:

P input 1	P$_{input\ 2}$	P$_{output}$
0	0	0
1	0	0
0	1	0
1	1	1

For OR Gate, the following logic applies:

P input 1	P$_{input\ 2}$	P$_{output}$
0	0	0
1	0	1
0	1	1
1	1	1

A fault tree analysis can be used to achieve the following:

- Define the undesired event to study
- Obtain an understanding of the system
- Construct the fault tree of the event
- Evaluating the event by a fault tree
- Control the hazards identified

Strengths of FTA are:

- It provides a useful graphical representation of hierarchical analysis of failure modes, since it provides a map that can help to understand the logic of the failure concerned.
- Can be used for diagnostics and fault finding. This is because it follows a logical and systematic process of breaking down a complex problem into its root causes.
- It can be quantified although it is inherently a qualitative analytic tool. It allows qualitative analysis to be performed i.e. the component failure combinations which result in the top event or system failure occurring. It allows quantitative analysis to be performed i.e. the probability of system failure, frequency of failure, system downtime and expected number of failures in x years.
- Helps to focus attention on the top event, a certain critical failure mode or an event, and then cascades deductively.
- It helps to identify the items of the system which are important the particular failure being studied.
- It provides ways of analysing human, software, procedural, and other factors in addition to the traditional physical parts of the system.
- A number of commercially available computer packages are now also available which provide a more efficient method of analysis using a Binary Decision Diagram.

Limitations of FTA are;

- If the system being analysed is complex, then the use of FTA for determining the order in which components fail becomes very difficult. In this case Event Tree Analysis would be more appropriate.
- Also if there are dependencies between components within a system; situations where standby redundancy, common cause failures or multiple component states are encountered then Markov Analysis should be used.
- Deterministic analysis becomes more difficult as system increases in size and complexity, or if component failure and repair distributions do not have constant failure or repair rates. In this case Simulation techniques should be used.

Reliability Block Diagrams (RBD)

Reliability block diagrams are logic diagrams arranged to indicate which combinations of component failures result in the failure of the system or which combination of properly working components keep the system functioning (Distefano, and Puliafito, 2007). Each block in the RBD represents an actual functioning component. Any failure is represented by removing the block from the diagram. If the connection between input and output is interrupted the system fails, however if only one path remains active from input to output the system is functional (Distefano, and Puliafito, 2007). Blocks could be arranged in two formats either in parallel or series, these different sections and connections could be all summarized into a single block and its reliability could be calculated using the series-parallel equations.

RBD's could be used to denote which combinations of component functionality that leads to stated system performance levels. Unlike FTA, RBD is a success oriented model; it looks at the combinations that will lead to system functionality. On the other hand in a fault tree, interest is in the system failure combinations (Xing and Amari, 2008). When the main concern is safety, it is recommended to begin by constructing a fault tree as this will make it possible to reveal more potential failures since the approach is failure oriented (Xing and Amari, 2008). This fault tree may be later converted into a reliability block diagram, and if an RBD is developed first, it is also possible to convert it to a fault tree in most cases. The conversion from a fault tree into a RBD begins from the top event of the fault tree, it is done by replacing each gate sequentially until the whole RBD is constructed, an AND logic gate is represented by a parallel structure whose elements are the inputs to the logic gate in the fault tree. An OR logic gate is represented by a series structure whose elements are the inputs to the logic gate in the fault tree. The blocks in a RBD indicate that components represented by blocks are working, while in a fault tree basic events indicate failure of components. If a conversion is required from a reliability block diagram into a fault tree analysis, a parallel structure is replaced by a fault tree section with all the inputs connected through an AND logic gate and a series structure is replaced by a fault tree with all the inputs connected by an OR logic gate (Xing and Amari, 2008).

Strengths of RBD's are;

- They allow for quantitative analysis to be performed
- Calculations for quantitative analysis are quite straight forward
- The interconnections within the RBD symbolise the way in which the system functions as required and is determined by the reliability dependencies
- Also this two-state assumption which is conservative makes reliability evaluations desirably cautious when safety assessments are involved

Limitations of RBD's are;

- Number of components involved must be reasonable
- The complexity of component inter-dependence should be low
- The interconnections within the RBD do not necessarily show how the basic events/ component failures are physically connected
- Failure probabilities of components are statistically independent i.e. the occurrence of failure of any one component in a system in no way influences or induces the occurrence of failure in any of the other components.
- Flow of time and sequence of failures are not captured in the process.

Acknowledgement:

The authors are grateful to the reviewer for the comprehensive and thoughtful comments and suggestions.

References

1. Alvi and A.W. Labib, (2001). "Selecting Next Generation Manufacturing Paradigms – An AHP Based Criticality Analysis", *Proc. Of IMechE, Journal of Engineering Manufacture - Part B*, 2(5), 1773-1786.
2. Argote, L. 1999. *Organizational learning: Creating, retaining, and transferring knowledge*. Boston: Kluwer Academic.

3. Balogun, J. and G. Johnson (2004). "Organizational Restructuring and Middle Manager Sensemaking " *The Academy of Management Journal* **47**(4): 496-523.

4. Barach, P. and S. D. Small (2000). "Reporting and preventing medical mishaps: lessons from nonmedical near miss reporting systems." *British Medical Journal* **3**20: 5.

5. Baum, J. A., & Dahlin, K. B. 2007. Aspiration performance and railroads' patterns of learning from train wrecks and crashes. *Organization Science*, **18**: 368– 385.

6. Beck, T. E., D. A. Plowman. 2009. Experiencing rare and unusual events richly: The role of middle managers in animating and guiding organizational interpretation. *Organ. Sci.* **20**(5) 909–924.

7. BP. (2005), *"Fatal accident investigation report, Isomerization unit Explosion Final Report"*, available at:http://www.bp.com/ liveassets/bp_internet/us/bp_usenglish/STAGING/local_assets/ downloads/t/final_report.pdf/ (Last accessed 20 March 2010)

8. Carroll, J. S., J. W. Rudolph and S. Hatakenaka (2002), 'Learning from experience in high hazard organizations,' *Research in Organizational Behavior*, **24**, 87–137.

9. Chuang YT, Ginsburg L, Berta WB. 2007. Learning from preventable adverse events in health care organizations: development of a multilevel model of learning and propositions, *Health Care Manage Rev.***32**(4):330-40.

10. Chouhan, T. R. 2005. "The Unfolding of Bhopal Disaster." *Journal of Loss Prevention in the Process Industries* **18**: 4-8.

11. Christianson, M. K., M. T. Farkas, K. M. Sutcliffe, K. E. Weick 2009. Learning through rare events: Significiant interruptions at the Baltimore & Ohio Railroad Museum. Organ. Sci. **20**(5) 846–860.

12. *Columbia Accident Investigation Board (CAIB, 2005)*, http://www. nasa.gov/columbia/caib/html/start.html (accessed 12/03/2005).

13. Davidson, G. and A. W. Labib 2003. "Learning from failures: design improvements using a multiple criteria decision making process." *Journal of Aerospace Engineering, Proceedings of the Institution of Mechanical Engineers Part G* **217**(1): 207-216.

14. Denrell, J. 2003. 'Vicarious learning, undersampling of failure, and the myths of management,' *Organization Science*, **14**(3), 227–243.

15. Desai, V. 2008. 'Constrained growth: How experience, legitimacy, and age influence risk taking in organizations,' *Organization Science*, **19**, 594–608.

16. Desai, V. 2010. Do organizations have to change to learn? Examining the effects of technological change and learning from failures in the natural gas distribution industry, *Industrial and Corporate Change*, **19** (3), pp. 713–739.

17. Distefano, S., Puliafito, A., (2007), *"Dynamic Reliability Block Diagrams VS Dynamic Fault Trees"*, in Reliability and Maintainability Symposium, 2007. RAMS '07. Annual, Orlando, Florida, pp. 71-76

18. Fernandez, O., A.W. Labib, R. Walmsley, D.J. Petty, "A Decision Support Maintenance Management System: Development and Implementation", *International Journal of Quality and Reliability Management*, Vol 20, No 8, pp 965-979, 2003.

19. Greening, D. W. and B. Gray (1994), 'Testing a model of organizational response to social and political issues,' *Academy of Management Journal*, 37(3), 467−498.

20. Haunschild, P. R., and Rhee, M. 2004. The role of volition in organizational learning: The case of automotive product recalls. *Management Science*, 50: 1545−1560.

21. Haunschild, P. R., & Sullivan, B. N. 2002. Learning from complexity: Effects of prior accidents and incidents on airlines' learning. *Administrative Science Quarterly*, 47: 609−643.

22. Holmstrom,D., Altamirano,F., Banks,J., Joseph,G., Kaszniak,M., Mackenzie,C., Shroff,R., Cohen,H., Wallace PE,S., CSP. (2006), "CSB Investigation of the Explosions and Fire at the BP Texas City Refinery on March 23, 2005", *Process Safety Progress*, Vol.25 No.4, pp.345-349

23. Hopkins, A. (2008), *Failure to Learn*, CCH Australia Limited, Australia.

24. IAEA Report INSAG-7, *Chernobyl Accident; Updating of INSAG-1*, Safety series No.75- INSAG-7 IAEA; Vienna (1991).

25. IAEA, (2005), Chernobyl's Legacy: Health, Environmental and Socio-economic Impacts and Recommendations to the Governments of Belarus, the Russian Federation and Ukraine. *The Chernobyl Forum*: 2003−2005, Second revised version.

26. Kageyama, Y. (2010) *Toyota recalls about 92,000 cars in Japan*, Associated Press, Published Tuesday, July 06 (http://www.onlineathens.com/stories/070610/nat_665672553.shtml)

27. Khan, F. I. and P. R. Amyotte (2007). "Modeling of BP Texas City refinery incident." *Journal of Loss Prevention in the Process Industries* 20(4-6): 387-395.

28. Kim, J., and Miner, A. S. 2000. *Crash test without dummies: A longitudinal study of interorganizational learning from failure experience in the U.S. commercial banking industry, 1984−1998.* Academy of Management Best Paper Proceedings.

29. Kim, J. Y., & Miner, A. S. 2007. Vicarious learning from the failures and near-failures of others: Evidence from the U.S. commercial banking industry. *Academy of Management Journal,* 50: 687−714

30. Kletz, T, Learning from Accidents, 2001, Butterworth-Heinemann.
31. Krasich, M., (2000), *"Use of Fault Tree Analysis for Evaluation of System-Reliability Improvements in Design Phase"*, in Reliability and Maintainability Symposium, 2000 Proceeding Annual, Los Angeles, California, pp. 1-7
32. Labib, A.W. (1999) "The Millennium Problem Versus the Maintenance Problem*", Journal of Logistics Information Management)*, Vol 12, No 3, pp 254-259.
33. Labib, A.W., (2004) A Decision Analysis Model for Maintenance Policy Selection Using a CMMS, *Journal of Quality in Maintenance Engineering*; MCB Press; ISSN: 1355-2511; Vol 10, No 3, pp 191-202.
34. Labib, A.W. Computerised Maintenance Management Systems, in *"Complex Systems Maintenance Handbook"*, Edited by: K.A. H. Kobbacy and D.N.P. Murthy, Springer, ISBN 978-1-84800-010-0, 2008.
35. Lampel, J., Shamsie, J and Shapira, Z., 2009, Experiencing the Improbable: Rare Events and Organizational Learning, *Organization Science*, 20 (5), 835−845
36. Madsen, P. M. 2009. These lives will not be lost in vain: Organizational learning from disaster in U.S. coal mining. *Organization Science.* 20(5) 861−875.
37. Madsen, P. M. and V. Desai (2010). "Failing to Learn? The Effects of Failure and Success on Organizational Learning in the Global Orbital Launch Vehicle Industry." *The Academy of Management Journal* 53(3): 451 − 476.
38. McGrath, R. G. 1999. Falling forward: Real options reasoning and entrepreneurial failure. *Academy of Management Review*, 24: 13−30.
39. Morris, M. W., & Moore, P. C. 2000. The lessons we (don't) learn: Counterfactual thinking and organizational accountability after a close call. *Administrative Science Quarterly*, 45: 737−765.
40. Mogford, J. (2005). *Fatal Accident Investigation Report: Isomerization Unit Explosion Final Report, Texas City, Texas, USA.* (http://people.uvawise.edu/pww8y/Supplement/-ConceptsSup/Work/WkAccidents/BPTxCityFinalReport.pdf)
41. Myddelton, D.R., *They meant well: Government Project Disasters*, IEA, Institute of Economic Affairs, 2007.
42. Nonaka I. 1994. A Dynamic theory of organisational knowledge. *Organisation Science*, 5 (1): 14-37.
43. Pearson, C.M. and Clair, J.A. (1998), 'Reframing Crisis Management', *Academy of Management Review*, 23 (1), 59-76.

44. Pearson, C.M. and Mitroff, I.I. (1993), 'From Crisis Prone to Crisis Prepared: a Framework for Crisis Management', *Academy of Management Executive*, 7 (1), 48-59.

45. Pearson, C.M. and Clair, J.A. (1998), 'Reframing Crisis Management', *Academy of Management Review*, 23 (1), 59-76.

46. Pidgeon, N., O'Leary, M., (2000) "Man-made disasters: why technology and organizations (sometimes) fail", *Safety Science*, 34 (1-3), 15-30.

47. Rerup, C. 2009. Attentional triangulation: Learning from unexpected rare crises. *Organ. Sci.* 20(5) 876—893.

48. Saaty, T.L., 1988, *The Analytic Hierarchy Process*, Pergamon Press, New York.

49. Saunders, M., Lewis, P. and Thornhill, A. (2003). 3rd. *Research Methods for Business Students*, Prentice Hall.

50. Shrivastava, P. (1992), *Bhopal. Anatomy of a Crisis* (2nd Edition), Paul Chapman Publishing, London.

51. Shrivastava, P., Mitroff, I.I., Miller, D. and Miglani, A. (1988), 'Understanding Industrial Crises', *Journal of Management Studies*, Volume 25, Number 4, July, pp. 283-303.

52. Sitkin, S. 1992. Learning through failure: The strategy of small losses. In B. M. Staw & L. L. Cummings (Eds.), *Research in organizational behavior*, vol. 14: 231—266. Greenwich, CT: JAI Press.

53. Smallman, C, (1996) "Challenging the orthodoxy in risk management", *Safety Science*, 22 (1-3), 245-262.

54. Smith, M.S., (2003) *NASA's Space Shuttle Columbia: Synopsis of the Report of the Columbia Accident Investigation Board,* http://history.nasa.gov/columbia/Troxell/Columbia%20Web%20Site/Documents/Congress/CRS%20Summary%20of%20CAIB%20Report.pdf

55. Smith, D. (1990), 'Beyond Contingency Planning: Towards a Model of Crisis Management', *Industrial Crisis Quarterly*, 4 (4), 263-275.

56. Starbuck, W. H. 2009. Cognitive reactions to rare events: Perceptions, uncertainty, and learning. *Organ. Sci.* 20(5) 925—937.

57. Stead, E., and Smallman, C., (1999), "Understanding Business Failure: Learning and Un-Learning From Industrial Crises", *Journal of Contingencies and Crisis Managemen*, 7 (1), 1-18.

58. Titanic (2000) *The Sinking of the Titanic, 1912, Eye Witness to History* www.eyewitnesstohistory.com (2000), accessed July, 2010.

59. Toft, B., and S. Rynolds, *Learning from disasters*, Perpetuity Press, 1997.

60. Towill, D. R. 'Management Theory': is it of any practical use? Or how does a fad become a paradigm? *Engineering Management Journal*, June 1999, **21**(2), 111-121.

61. Tsoukas, H. and R. Chia (2002). "On Organizational Becoming: Rethinking Organizational Change" *Organization Science* **13**(5): 16.

62. Turner, B A., (1978) Man-made disasters, Wykeham science series;, 1978, 254p

63. Turner, B.A. and Pidgeon, N.F. (1997), *Man-Made Disasters* (2nd Edition), Butterworth-Heinemann, London.

64. van de Ven, A. H. and M. S. Poole (1995). "Explaining Development and Change in Organizations" *The Academy of Management Review* **20**(3): 510-541.

65. Vaidogas, E.R., and Juocevičius, V. (2008) Sustainable development and major industrial accidents: the beneficial role of risk-oriented structural engineering, Technological and economic development of Economy, *Baltic Journal on Sustainability* **14**(4): 612−627.

66. Vaughan, D. 1996. *The Challenger launch decision.* Chicago: University of Chicago Press.

67. Vaughan, D. 2005. System effects: On slippery slopes, repeating negative patterns, and learning from mistake? In M. Farjoun, W. Starbuck (Eds). *Organization at the Limit: NASA and the Columbia Disaster.* Blackwell, Oxford, UK.

68. Weir D. *The Bhopal Syndrome: Pesticides, Environment and Health.* San Francisco; 1987.

69. World Nuclear Association; *Chernobyl Accident, revisited* 2009. http://www.world-nuclear.org

70. James Womack, Daniel T. Jones and Daniel Roos, *The Machine that Changed the World*, 1990

71. Wood, W., S. Lundgren, J. A. Ouellette, S. Busceme, T. Blackstone. 1994. Minority influence: A meta-analytic review of social influence processes. *Psychological Bulletin*, **115**, 323-345.

72. Xing,L., Amari,S.V. (2008), "Fault Tree Analysis", in Misra, K.B.(Ed.), *Handbook of Per formability Engineering*, Springer London, London, pp. 595-620.

73. Zollo, M. 2009. Superstitious learning with rare strategic decisions: Theory and evidence from corporate acquisitions. *Organ. Sci.* **20**(5) 894−908.

References

Abdi, M.R., Labib, A.W., 2003. A design strategy for reconfigurable manufacturing systems (RMSs) using the analytical hierarchical process (AHP): a case study. Int. J. Prod. Res. 41 (10), 2273–2301.

Ahearne, J.F., 1987. Nuclear power after Chernobyl. Science 236 (4802), 673–679.

Alvi, Labib, A.W., 2001. Selecting next generation manufacturing paradigms—an AHP based criticality analysis. Proc. I. Mech. E J. Eng Manuf.—Part B 2 (5), 1773–1786.

American Society of Civil Engineers (ASCE), 2007. The New Orleans Hurricane protection system: what went wrong and why (online). Available from: <http://www.asce.org/files/pdf/ERPreport.pdf> (accessed 31.03.12).

Apostolakis, G.E., 2004. How useful is quantitative risk assessment? Risk Anal. 24 (3), 515–520.

Areva, The Fukushima Daiichi Incident, 2011. Presentation can be found at: <hps.org/documents/areva_japan_accident_20110324.pdf>.

Argote, L., 1999. Organizational Learning: Creating, Retaining, and Transferring Knowledge. Kluwer Academic, Boston, MA.

Argote, L., Epple, D., 1990. Learning curves in manufacturing. Science 23, 920–924.

Aslam-Zainudeen, Labib, A., 2011. Practical application of the decision making grid (DMG). J. Qual. Main. Eng. 17 (2), 138–149, MCB Press; ISSN: 1355-2511.

Attard, K., 2013. People's Perception on Cruise Liners Disasters, Final Year Project. Institute of Tourism, Travel and Culture, University of Malta, Malta.

Aven, T., 1992. Reliability and Risk Analysis. Elsevier, London.

Aven, T., 2011. On some recent definitions and analysis frameworks for risk, vulnerability, and resilience. Risk Anal. 31 (4), 515–522.

Baker Panel. 2007. The Report of the BP U.S. Refineries Independent Safety Review Panel. The Baker Panel, Washington, DC.

Balogun, J., Johnson, G., 2004. Organizational restructuring and middle manager sensemaking. Acad. Manage. J. 47 (4), 496–523.

Barach, P., Small, S.D., 2000. Clinical review reporting and preventing medical mishaps: lessons from non-medical near miss reporting systems. Br. Med. J., 759–763.

Baum, J.A., Dahlin, K.B., 2007. Aspiration performance and railroads' patterns of learning from train wrecks and crashes. Organ. Sci. 18, 368–385.

BBC News, 2011. How does Fukushima differ from Chernobyl? Available from: <http://www.bbc.co.uk/news/world-asia-pacific-13050228> (accessed 16.12.11).

Beck, T.E., Plowman, D.A., 2009. Experiencing rare and unusual events richly: the role of middle managers in animating and guiding organizational interpretation. Organ. Sci. 20 (5), 909–924.

Bier, V.M., Haimes, Y.Y., Lambert, J.H., Matals, N.C., Zimmerman, R., 1999. A survey of approaches for assessing and managing the risk of extremes. Risk Anal. 19 (1), 83–94.

BP, 2005. Fatal accident investigation report, Isomerization Unit Explosion Final Report. Available from: <http://www.bp.com/liveassets/bp_internet/us/bp_usenglish/STAGING/local_assets/downloads/t/final_report.pdf/> (accessed 20.03.10).

BP Commission Report, 2011. National Commission on the BP Deepwater Horizon Oil Spill & Offshore Drilling, Deepwater: The Gulf Oil Disaster and the Future of Offshore Drilling 1. Available from: <http://www.oilspillcommission.gov/sites/default/files/documents/DEEPWATER_ReporttothePresident_FINAL.pdf> (accessed 12.11.13).

BP Texas City Report, 2007. Investigation Report, Refinery Explosion and Fire, BP Texas City, Texas. U.S. Chemical Safety and Hazard Investigation Board, Washington, DC.

Bradley, J.R., Guerrero, H.H., 2011. An alternative FMEA method for simple and accurate ranking of failure modes. Decis. Sci. 42 (3), 743–771.

Buljan, A., Shapira, Z., 2005. Attention to production schedule and safety as determinants of risk-taking in NASA's decision to launch the Columbia shuttle. In: Starbuck, W.H., Farjoun, M. (Eds.), Organization at the Limit: Lessons from the Columbia Disaster. Blackwell Publishing, New York.

Buncefield Major Incident Investigation Board, 2008. The Buncefield Incident 11 December 2005: The Final Report of Major Incident Investigation Board, vols. 1–2. HSE Books, London.

Bureau d'Enquetesetd'Analysespur la Securitedel'AviationCivile, 2002. Accident on 25 July 2000 at La Patted'Oie in Gonesse (95) to the Concorde registered F-BTSC operated by Air France (English Translation). Ministere de l'Equipement des Transports et du Logement, France.

Burhanuddin, M.A., 2007. An application of decision making grid to improve maintenance strategies in small and medium industries. Proceedings of the Second IEEE Conference on Industrial Electronics and Applications, Singapore, pp. 455–460.

Carroll, J.S., Rudolph, J.W., Hatakenaka, S., 2002. Learning from experience in high hazard organizations. Res. Organ. Behav. 24, 87–137.

Cartwright, R., Baird, C., 1999. The Development and Growth of the Cruise Industry, first ed. Butterworth-Heinemann, London.

Cassels, J., 1993. The Uncertain Promise of Law: Lessons from Bhopal. University of Toronto Press, Toronto.

Chase, R.B., Stuart, D.M., 1994. Make your service fail-safe. Sloan Manage. Rev. 35 (3).

Chemical Safety Board, 2010. BP/Transocean Deepwater Horizon Oil Rig Blowout.

Chouhan, T.R., 2005. The unfolding of Bhopal disaster. J. Loss Prev. Process Ind. 18, 4–8.

Christianson, M.K., Farkas, M.T., Sutcliffe, K.M., Weick, K.E., 2009. Learning through rare events: significant interruptions at the Baltimore & Ohio Railroad Museum. Organ. Sci. 20 (5), 846–860.

Chuang, Y.T., Ginsburg, L., Berta, W.B., 2007. Learning from preventable adverse events in health care organizations: development of a multilevel model of learning and propositions. Health Care Manage. Rev. 32 (4), 330–340.

Columbia Accident Investigation Board (CAIB), 2003. Columbia Accident Investigation Board Report, vols. 1–6. Apogee Books, Burlington, Ontario, Canada.

Columbia Accident Investigation Board (CAIB), 2005. Available from: <http://www.nasa.gov/columbia/caib/html/start.html> (accessed 12.03.05).

Cox, S., Flin, R., 1998. Safety culture: philosopher's stone or man of straw? Work Stress 12 (3), 189–201.

Crowther, K.G., Haimes, Y.Y., Taub, G., 2007. Systemic valuation of strategic preparedness through application of the inoperability input-output model with lessons learned from Hurricane Katrina. Risk Anal. 27 (5), 1345–1364.

Cullen, The Hon. Lord W. Douglas. 1990. The Public Inquiry into the Piper Alpha Disaster. H.M. Stationery Office, London

Cyert, R.M., March, J.G., 1963. A Behavioral Theory of the Firm. Oxford University Press, New York, NY.

D'Adderio, L., 2008. The performativity of routines: theorising the influence of artefacts and distributed agencies on routines dynamics. Res. Policy 37 (5), 769–789.

Dale, B.G., Shaw, P., 1990. Failure mode and effect analysis: a study of its use in the motor industry. An Occasional Paper 8904, UMIST, Manchester.

Davidson, G., Labib, A.W., 2003. Learning from failures: design improvements using a multiple criteria decision making process. J. Aerosp. Eng. Proc. Inst. Mech. Eng. Part G 217 (1), 207–216.

Denrell, J., 2003. Vicarious learning, undersampling of failure, and the myths of management. Organ. Sci. 14 (3), 227–243.

Desai, V., 2008. Constrained growth: how experience, legitimacy, and age influence risk taking in organizations. Organ. Sci. 19, 594–608.

Desai, V., 2010. Do organizations have to change to learn? Examining the effects of technological change and learning from failures in the natural gas distribution industry. Ind. Corporate Change 19 (3), 713–739.

DHSG, 2011. Final report on the investigation of the Macondo well blowout. Deepwater Horizon study group.

Dijkstra, A., 2006. Safety management in airlines. In: Hollnagel, E., Woods, D.D., Leveson, N. (Eds.), Resilience Engineering Concepts and Precepts. Ashgate, England.

Dillon, R.L., Tinsley, C.H., 2008. How near-misses influence decision making under risk: a missed opportunity for learning. Manage. Sci. 54 (8), 1425–1440.

Dillon, R.L., Tinsley, C.H., Madsen, P.M., Rogers, E.W., 2012. Organizational correctives for improving recognition of near-miss events (working paper).

Division of the History of Technology, Transportation Collections, National Museum of American History, in cooperation with the Public Inquiry Mail Service, Smithsonian Institution, The Titanic. May 1997. Available from: <http://www.si.edu/resource/faq/nmah/titanic.htm> Washington, DC, Smithsonian Institution.

Dobias, A.P., 1990. Designing a mouse trap using the analytic hierarchy process and expert choice. Eur. J. Oper. Res. 48, 57–65.

DoE, 1999. Conducting Accident Investigations DOE Workbook, Revision 2. US Department of Energy, Washington, DC.

Einestein, A., Infeld, L., 1938. The Evolution of Physics. Simon and Schuster, New York, NY.

Elkind, P., Whitford, D., Burke, B., January 24, 2011. BP: 'An accident waiting to happen', CNN Money.

Farjoun, M., 2010. Beyond dualism: stability and change as a duality. Acad. Manage. Rev. 35 (2), 202–225.

Farquharson, J., McDuffee, J., Seah, A.K., Matsumoto, T., 2002. FMEA of marine systems: moving from prescriptive to risk-based design and classification. Annu. Reliab. Maintainability Symp. Proc., 165–172.

Feldman, M.S., 2000. Organizational routines as a source of continuous change. Organ. Sci. 11 (6), 611–629.

Feldman, M.S., Pentland, B.T., 2003. Reconceptualizing organizational routines as a source of flexibility and change. Adm. Sci. Q. 48 (1), 94–118.

Fernandez, O., Labib, A.W., Walmsley, R., Petty, D.J., 2003. A decision support maintenance management system: development and implementation. Int. J. Qual. Reliab. Manage. 20 (8), 965–979.

Flin, R., 2006. Erosion of managerial resilience: from VASA to NASA. In: Hollnagel, E., Woods, D.D., Leveson, N. (Eds.), Resilience Engineering Concepts and Precepts. Ashgate, England.

Flouron, A.C., 2011. Three Meta-Lessons Government and Industry Should Learn from the BP Deepwater Horizon Disaster and Why They Will Not. 38 (2), Boston College Environmental. Affairs Law Review.

Forman, E., Gass, S., 2001. The analytic hierarchy process—an exposition. Oper. Res. 49 (4), 469–486.

Fowler, T., 2010. Spill response compared to Apollo 13 effort, Houston Chronicle, 6 May.

Franceschini, F., Galetto, M., 2001. A new approach for evaluation of risk priorities of failure modes in FMEA. Int. J. Prod. Res. 39 (13), 2991–3002.

Golden, B., Wasil, E., et al., 1989. The Analytic Hierarchy Process: Applications and Studies. Springer-Verlag, Heidelberg.

Gould, S.J., 1981. The Mismeasure of Man. Norton, New York, NY.

Greening, D.W., Gray, B., 1994. Testing a model of organizational response to social and political issues. Acad. Manage. J. 37 (3), 467–498.

Griffis, F.H., 2007. Engineering failures exposed by Hurricane Katrina. Technol. Soc. 29, 189–195.

Hartmann, E.H., 1992. Successfully Installing TPM in a Non-Japanese Plant. TPM Press, Inc., New York, NY.

Haunschild, P.R., Rhee, M., 2004. The role of volition in organizational learning: the case of automotive product recalls. Manage. Sci. 50, 1545–1560.

Haunschild, P.R., Sullivan, B.N., 2002. Learning from complexity: effects of prior accidents and incidents on airlines' learning. Adm. Sci. Q. 47, 609–643.

Heinmann, L., 2005. Repeated failures in the management of high risk technologies. Eur. Manage. J. 23 (1), 105–117.

Ho, W., 2008. Integrated analytic hierarchy process and its applications—a literature review. Eur. J. Oper. Res. 186 (1), 211–228.

Hodge, B.K., 2010. Alternative Energy Systems and Applications. John Wiley & Sons, New Jersey, NJ, USA.

Hollnagel, E., Woods, D.D., Leveson, N., 2006. Resilience Engineering Concepts and Precepts. Ashgate, England.

Holmstrom, D., Altamirano, F., Banks, J., Joseph, G., Kaszniak, M., Mackenzie, C., et al., 2006. CSB investigation of the explosions and fire at the BP Texas City refinery on March 23, 2005. Process Saf. Prog. 25 (4), 345–349.

Holroyd, T., 2000. Acoustic Emission & Ultrasonics. Coxamoor Publishing Company, Oxford.

Hopkins, A., 2008. Failure to Learn, the BP Texas City Refinery Disaster. CCH Australia Limited, Sydney, NSW, Australia.

IAEA, 2005. Chernobyl's Legacy: Health, Environmental and Socio-economic Impacts and Recommendations to the Governments of Belarus, the Russian Federation and Ukraine. The Chernobyl Forum: 2003–2005, Second revised version.

IAEA Report INSAG-7, 1991. Chernobyl Accident; Updating of INSAG-1, Safety series No.75- INSAG-7 IAEA; Vienna.

International Nuclear Safety Advisory Group, INSAG-7, 1992. The Chernobyl Accident: Updating of INSAG-1. (75-INSAG-7). IAEA.

Ishizaka, A., Labib, A., 2009. Analytic Hierarchy process and expert choice: benefits and limitations. ORInsight 22 (4), 201–220, ISSN: 0953-5543.

Ishizaka, A., Labib, A., 2013. A hybrid and integrated approach to evaluate and prevent disasters. J. Oper. Res. Soc.

Ishizaka, A., Nemery, P., 2013. Multi-Criteria Decision Analysis Methods and Software. John Wiley & Sons, United Kingdom.

Ishizaka, A., Balkenborg, D., Kaplan, T., 2011. Does AHP help us make a choice? An experimental evaluation. J. Oper. Res. Soc. 62 (10), 1801–1812.

ISO:13702, 1999. Petroleum and natural gas industries—control and mitigation of fires and explosions on offshore production installations—requirements and guidelines. Geneva, International Organization for Standardization.

Jardine, A.K.S., Lin, D., Banjevic, D., 2006. A review on machinery diagnostics and prognostics implementing condition-based maintenance. Mech. Syst. Sig. Process. 20 (7), 1483–1510.

JNES, 2005. Outline of Safety Design—A Case of BWR, Long-Term Training Course Notes, Japan Nuclear Energy Safety.

Jonkman, S.N., Maaskant, B., Boyd, E., Levitan, M.L., 2009. Loss of life caused by the flooding of New Orleans after Hurricane Katrina: analysis of the relationship between flood characteristics and mortality. Risk Anal. 29 (5), 676–698.

Kageyama, Y., 2010. Toyota recalls about 92,000 cars in Japan. Associated Press, Published Tuesday, July 06. Available from: <http://www.onlineathens.com/stories/070610/nat_665672553.shtml>.

Kahneman, D., 2012. Thinking, Fast and Slow. Penguin, England.

Kaplan, S., Garrick, B.J., 1981. On the quantitative definition of risk. Risk Anal. 1 (1), 11–27.

Kashyap, A., et al., 2011. The Economic Consequences of the Earthquake in Japan, Web Publication. Freakonomics Web, Chicago, USA.

Khan, F.I., Amyotte, P.R., 2007. Modeling of BP Texas City refinery incident. J. Loss Prev. Process Ind. 20 (4–6), 387–395.

Kim, J., Miner, A.S., 2000. Crash test without dummies: a longitudinal study of interorganizational learning from failure experience in the U.S. commercial banking industry, 1984–1998. Academy of Management Best Paper Proceedings.

Kim, J.Y., Miner, A.S., 2007. Vicarious learning from the failures and near-failures of others: evidence from the U.S. commercial banking industry. Acad. Manage. J. 50, 687–714.

Kletz, T., 2001. Learning from Accidents. Butterworth-Heinemann, Oxford, UK.

Kletz, T., 2009. What Went Wrong? Case Histories of Process Plant Disasters and How They Could Have Been Avoided, fifth ed. Elsevier Inc., ISBN: 978-1-85617-531-9.

Koestler, A., 1989. The Ghost in the Machine. Arkana Books, London.

Krolicki, K., DiSavino, S., Fuse, T., 2011. Engineers knew Tsunami Could Overwhelm Fukushima Plant. Insur. J. Available from: <http://www.insurancejournal.com/news/international/2011/03/30/192204.htm>.

Kumar, S., Vaidya, O., 2006. Analytic hierarchy process: an overview of applications. Eur. J. Oper. Res. 169 (1), 1–29.

Labib, A., Champaneri, R., 2012. The Bhopal disaster—learning from failures and evaluating risk. J. Main. Asset Manage. 27 (3), 41–47.

Labib, A., Read, M., Gladstone-Millar, C., Tonge, R., Smith, D., 2013. Formulating a higher institution educational strategy using operational research approaches, Studies in Higher Education.

Labib, A.W., 1998. World class maintenance using a computerised maintenance management system. J. Qual. Main. Eng. 4 (1), 66–75, MCB Press, ISSN: 1355-2511.

Labib, A.W., 1999. The millennium problem versus the maintenance problem. J. Logist. Inf. Manage. 12 (3), 254–259.

Labib, A.W., 2004. A decision analysis model for maintenance policy selection using a CMMS. J. Qual. Main. Eng. 10 (3), 191–202, MCB Press.

Labib, A.W., 2008. Computerised maintenance management systems. In: Kobbacy, K.A.H., Murthy, D.N.P. (Eds.), Complex Systems Maintenance Handbook. Springer, Berlin, ISBN 978-1-84800-010-0.

Labib, A.W., 2011. A supplier selection model: a comparison of fuzzy logic and the analytic hierarchy process. Int. J. Prod. Res. (IJPR) 49 (21).

Labib, A., Read, M., 2013. Not just rearranging the deckchairs on the Titanic: learning from failures through risk and reliability analysis. Saf. Sci. 51, 397–413.

Labib, A.W., Shah, J., 2001. Management decisions for a continuous improvement process in industry using the analytical hierarchy process. J. Work Study 50 (5), 189–193, ISSN 0043-8022.

Labib, A.W., O'Connor, R.F., Williams, G.B., 1996. Formulation of an appropriate maintenance strategy using multiple criteria decision making. Main. J. 11 (2), 14–21, ISSN:0952-2110.

Labib, A.W., O'Connor, R.F., Williams, G.B., 1997. Deriving a maintenance strategy through the application of a multiple criteria decision making methodology, ISSN:0075-8442, ISBN:3-540-62097-4.In: Fandel, G., Gal, T. (Eds.), Lecture Notes in Economics and Mathematical Systems, vol. 448. Springer-Verlag, Berlin.

Labib, A.W., Williams, G.B., O'Connor, R.F., 1998a. An intelligent maintenance model (System): an application of A.H.P. and a fuzzy logic rule-based controller. J. Oper. Res. Soc. 9 (7), 745–757, ISSN: 0160-5682.

Labib, A.W., O'Connor, R.F., Williams, G.B., 1998b. An effective maintenance system using the analytic hierarchy process. J. Integr. Manuf. Syst. 9 (2), 87–98, ISSN: 0957-6061.

Lampel, J., 2006. Rules in the shadow of the future: prudential rule making under ambiguity in the aviation industry. Inter. Relat. 20 (3), 343–349.

Lampel, J., Shamsie, J., Shapira, Z., 2009. Experiencing the improbable: rare events and organizational learning. Organ. Sci. 20 (5), 835–845.

Latour, B., 1991. Technology is society made durable. In: Law, J. (Ed.), A Sociology of Monsters. Essays on Power, Technology and Domination. Routledge, London, pp. 103–131.

Le Coze, J., 2008. Disasters and organisations: from lessons learnt to theorising. Saf. Sci. 46, 132–149.

Leveson, N., 1995. Safeware: System Safety and Computers. Addison-Wesley Publishers, Cambridge, Massachusetts.

Leveson, N., 2004. A new accident model for engineering safer systems. Saf. Sci. 42/4, pp. 237–270.

Leveson, N., Cutcher-Gershenfeld, J., Carroll, J.S., Barrett, B., Brown, A., Dulac, N., et al., 2005. Systems approaches to safety: NASA and the space shuttle disasters. In: Starbuck, W.H., Farjoun, M. (Eds.), Organization at the Limit: Lessons from the Columbia Disaster. Blackwell Publishing, Malden, MA, USA.

Leveson, N., Dulac, N., Marais, K., Carroll, J., 2009. Moving beyond normal accidents and high reliability organizations: a systems approach to safety in complex systems. Organ. Stud. 30 (2–3), 227–249.

Levinthal, D.A., Rerup, C., 2006. Crossing an apparent chasm: bridging mindful and less mindful perspectives on organizational learning. Organ. Sci. 17 (4), 502–513.

Levitt, B., March, J.G., 1988. Organizational learning. Ann. Rev. Sociol. 14, 319–340.

Liberatore, M., Nydick, R., 2008. The analytic hierarchy process in medical and health care decision making: a literature review. Eur. J. Oper. Res. 189 (1), 194–207.

Madsen, P.M., 2009. These lives will not be lost in vain: organizational learning from disaster in U.S. coal mining. Organ. Sci. 20 (5), 861–875.

Madsen, P.M., Desai, V., 2010. Failing to learn? The effects of failure and success on organizational learning in the global orbital launch vehicle industry. Acad. Manage. J. 53 (3), 451–476.

Mahler, J.G., Casamayou, M.H., 2009. More to Learn from NASA about learning, unlearning, and forgetting. Organizational Learning at NASA: The Challenger & Columbia Accidents. Georgetown University Press, Washington, DC.

March, J.G., 1991. Exploration and exploitation in organizational learning. Organ. Sci. 2, 71–87.

March, J.G., Simon, H.A., 1958. Organizations. Wiley, New York, NY.

McDonald, N., 2006. Organizational resilience and industrial risk. In: Hollnagel, E., Woods, D.D., Leveson, N. (Eds.), Resilience Engineering Concepts and Precepts. Ashgate, England.

McGrath, R.G., 1999. Falling forward: real options reasoning and entrepreneurial failure. Acad. Manage. Rev. 24, 13–30.

MIL-STD-1629A, 1980. Procedures for Performing a Failure Mode, Effects, and Criticality Analysis.

Miller, K.D., Pentland, B.T., Choi, S., 2012. Dynamics of performing and remembering organizational routines. J. Manage. Stud. 49, 8.

Mogford, J., 2005. Fatal Accident Investigation Report: Isomerization Unit Explosion Final Report, Texas City, Texas, USA. Available from: <http://people.uvawise.edu/pww8y/Supplement/ConceptsSup/Work/WkAccidents/BPTxCityFinalReport.pdf> (accessed 23.11.11).

Mohr, R.R., 2002. Failure Modes and Effects Analysis. Jacobs Engineering Group, Pasadena, CA.

Morris, M.W., Moore, P.C., 2000. The lessons we (don't) learn: counterfactual thinking and organizational accountability after a close call. Adm. Sci. Q. 45, 737–765.

Morris-Suzuki, T., Boilley, D., McNeill, D., Gundersen, A., 2012. Lessons from Fukushima. Greenpeace International, Netherlands.

Moubray, J., 1991. Reliability Centred Maintenance. Butterworth-Heinmann Ltd, Oxford.

Nakajima, S., 1988. Total Productive Maintenance. Productivity Press, Chicago, IL.

Narduzzo, A., Rocco, E., Warglien, M., 2000. Talking about routines in the field. In: Dosi, G., Nelson, R., Winter, S. (Eds.), The Nature and Dynamics of Organizational Capabilities. Oxford University Press, New York.

Nelson, R.R., Winter, S.G., 1982. An Evolutionary, Theory of Economic Change. Harvard University, Cambridge, MA.

Netherton, D., 2000. RCM standard. Maintenance Asset Manage. 15, 12−20.

Nolan, F., Heap, H., 1979. Reliability Centred Maintenance, National Technical Information Service Report, # A066-579.

Nonaka, I., Takeuchi, H., 1995. The Knowledge-Creating Company. Oxford University Press, New York, NY.

Ocasio, 2005. The opacity of risk: language and the culture of safety in NASA's space shuttle program. In: Starbuck, W.H., Farjoun, M. (Eds.), Organization at the Limit: Lessons from the Columbia Disaster. Blackwell Publishing, Malden, MA, USA.

O'Connor, P.D., 1990. Practical Reliability Engineering. John Wiley & Sons, New York.

Omkarprasad, V., Sushil, K., 2006. Analytic hierarchy process: an overview of applications. Eur. J. Oper. Res. 169 (1), 1−29.

Pate-Cornell, M.E., 1993. Learning from the Piper Alpha accident: a postmortem analysis of technical and organizational factors. Risk Anal. 13 (2), 215−232.

Pearson, C.M., Clair, J.A., 1998. Reframing crisis management. Acad. Manage. Rev. 23 (1), 59−76.

Pearson, C.M., Mitroff, I.I., 1993. From crisis prone to crisis prepared: a framework for crisis management. Acad. Manage. Exec. 7 (1), 48−59.

Pentland, B.T., Feldman, M.S., Becker, M.C., Liu, P., 2012. Dynamics of organizational routines: a generative model. J. Manage. Stud. 49, 8.

Perin, C., 2005. Shouldering Risks: The Culture of Control in the Nuclear Power Industry. Princeton University Press, Princeton.

Perrow, C., 1984. Normal Accidents: Living with High-Risk Technologies. Basic Books, New York, NY.

Petroski, H., 2006. Success Through Failure: The Paradox of Design. Princeton University Press, Princeton, NJ.

Pfotenhauer, S., KrishanuSaha, J., Jasanoff, S., Learning from Fukushima, World Nuclear News, Events Scale. Available from: <www.world-nuclear-news.org> (accessed 13.10.13).

Pidgeon, N., 1998. Safety culture: key theoretical issues. Work Stress 12 (3), 202.

Pidgeon, N., O'Leary, M., 2000. Man-made disasters: why technology and organizations (sometimes) fail. Saf. Sci. 34 (1−3), 15−30.

Pillay, A., Wang, J., 2003. Modified failure mode and effects analysis using approximate reasoning. Reliab. Eng. Syst. Saf. 79 (1), 69−85.

Presidential Commission, 1986. Report of the Presidential Commission on the Space Shuttle Challenger Accident. US Government Printing Office, Washington, DC.

Price, C.J., Taylor, N.S., 2002. Automated multiple failure FMEA. Reliab. Eng. Syst. Saf. 76 (1), 1−10.

Rasmuseen, J., 1997. Risk management in a dynamic society: a modelling problem. Saf. Sci. 27/2, 183−213.

Reason, J.T., 1987. The Chernobyl errors. Bull. Br. Psychol. Soc. 4, 201−206.

Reason, J.T., 1997. Managing the Risks of Organizational Accidents. Ashgate, Aldershot, Hants, England; Brookfield, Vermont, USA.

Rerup, C., 2009. Attentional triangulation: learning from unexpected rare crises. Organ. Sci. 20 (5), 876−893.

Rerup, C., Feldman, M.S., 2011. Routines as a source of change in organizational schemata: the role of trial-and-error learning. Acad. Manag. J. 54 (3), 577–610.

Rijpma, J.A., 1997. Complexity, tight-coupling and reliability: connecting normal accidents theory and high reliability theory. J. Contingencies Crisis Manage. 15 (15).

Rigzone, 2010. How do semisubmersibles work? Available from: <http://www.rigzone.com/training/insight.asp?insight_id = 338&c_id = 24> (accessed 23.11.10).

Rochlin, G.I., La Porte, T.R., Roberts, K.H., 1987. The self-designing high reliability organization. Reprinted in Naval War College Review 1998; 51 (3), 17.

Rogers Commission, Presidential Commission on the Space Shuttle Challenger Accident, 1986. Report of the Presidential Commission on the Space Shuttle Challenger Accident. Washington, DC: GPO. Available from: <http://science.ksc.nasa.gov/shuttle/missions/51-l/docs/rogers-commission> (accessed 01.02.08).

Saaty, T., 1972. An eigenvalue allocation model for prioritization and planning. Working paper, Energy Management and Policy Center, University of Pennsylvania.

Saaty, T., Forman, E., 1992. The Hierarchon: A Dictionary of Hierarchies. RWS Publications, Pittsburgh.

Saaty, T.L., 1977. A scaling method for priorities in hierarchical structures. J. Math. Psychol. 15 (1), 57–68.

Saaty, T.L., 1980. The Analytic Hierarchy Process. McGraw-Hill International, New York. NY.

Saaty, T.L., 1990. How to make a decision: the analytic hierarchy process. Eur. J. Oper. Res. 48, 9–26.

Saaty, T.L., 1996. Decision Making with Dependence and Feedback: The Analytic Network Process. RWS Publications, Pittsburgh, PA.

Saleh, J.H., Pendley, C.C., 2012. From learning from accidents to teaching about accident causation and prevention: multidisciplinary education and safety literacy for all engineering students. Reliab. Eng. Syst. Saf. 99, 105–113.

Saleh, J.H., Marais, K.B., Bakolas, E., Cowlagi, R.V., 2010. Highlights from the literature on system safety and accident causation: review of major ideas, recent contributions, and challenges. Reliab. Eng. Syst. Saf. 95 (11), 1105–1116.

Salvato, C., Rerup, C., 2011. Beyond collective entities: multilevel research on organizational routines and capabilities. J. Manage. 37 (2), 468–490.

Saunders, M., Lewis, P., Thornhill, A., 2003. Research Methods for Business Students, third ed. Prentice Hall, Harlow.

Select Bipartisan Committee to Investigate the Preparation for and Response to Hurricane Katrina, 2006. A Failure of Initiative. U.S. Government Printing Office, Washington DC, USA.

Seneviratne, G., August 2011. Director General to Set Out Post-Fukushima Action Plan, IAEA Ministerial Conference, Nuclear News.

Shim, J., 1989. Bibliography research on the analytic hierarchy process (AHP). Socioecon. Plann. Sci. 23 (3), 161–167.

Shrivastava, P., Mitroff, I.I., Miller, D., Miglani, A., 1988. Understanding industrial crises. J. Manage. Stud. 25 (4), 283–303.

Sitkin, S., 1992. Learning through failure: the strategy of small losses. In: Staw, B.M., Cummings, L.L. (Eds.), Research in Organizational Behavior, vol. 14. JAI Press, Greenwich, CT, pp. 231–266.

Sklet, S., 2006. Safety barriers: definition, classification, and performance. J. Loss Prev. Process Ind. 19, 494–506.

Slack, N., Chambers, S., Johnston, R., 2009. Operations & Process Management, fifth ed. Prentice Hall, Harlow, UK.

Smallman, C., 1996. Challenging the orthodoxy in risk management. Saf. Sci. 22 (1–3), 245–262.

Smith, M.S., 2003. NASA's Space Shuttle Columbia: Synopsis of the Report of the Columbia Accident Investigation Board. Available from: <http://history.nasa.gov/columbia/Troxell/Columbia%20Web%20Site/Documents/Congress/CRS%20Summary%20of%20CAIB%20Report.pdf> (accessed 28.3.09.).

Snell Dr, V.G., Howieson, J.Q., 1991. Chernobyl—A Canadian Perspective. Atomic Energy of Canada Ltd, Ottawa, Canada.

Sorensen, J.N., 2002. Safety culture: a survey of the state of the art. Reliab. Eng. Syst. Saf. 76 (2), 189–204.

Stamatis, DH, 1995. Failure Mode and Effect Analysis: FMEA from Theory to Execution. ASQ Press, Milwaukee, WI.

Starbuck, W.H., 2009. Cognitive reactions to rare events: perceptions, uncertainty, and learning. Organ. Sci. 20 (5), 925–937.

Starbuck, W.H., Milliken, F.J., 1988. *Challenger*: fine-tuning the odds until something breaks. J. Manage. Stud. 25, 319–340.

Starbuck, W.H., Farjoun, M. (Eds.), 2005. Organization at the Limit: Lessons from the Columbia Disaster. Blackwell Publishing, Malden, MA, USA.

Stead, E., Smallman, C., 1999. Understanding business failure: learning and un-learning from industrial crises. J. Contingencies Crisis Manage. 7 (1), 1–18.

Stellman, J.M., 1998. fourth ed. Encyclopedia of Occupational Health and Safety, vol. 2. International Labour Organisation, Geneva.

Sunstein, C.R., 2005. The Laws of Fear; Beyond the Precautionary Principle. Cambridge University Press, New York, NY.

Svenson, O., 1991. The accident evolution and barrier function (AEB) model applied to incident analysis in the processing industries. Risk Anal. 11 (3), 499–507.

Tahir, Z., Prabuwono, A.S., Aboobaider, B.M., 2008. Maintenance decision support system in small and medium industries: an approach to new optimization model. Int. J. Comput. Sci. Netw. Secur. 8 (11), 155–162.

Talbert, M.L., Balci, O., Nance, R.E., 1996. Application of the Analytic Hierarchy Process to Complex System Design Evaluation. Virginia Polytechnic Institute and State University.

Taleb, N.N., 2010. The Black Swan: The Impact of the Highly Improbable. Penguin Books, New York.

Tamuz, M., 2000. Defining away dangers: a study in the influences of managerial cognition on information systems. In: Lant, T., Shapira, Z. (Eds.), Organizational Cognition: Computation and Interpretation. Lawrence Erlbaum, Mahwah, NJ, pp. 157–184.

Temiz, N., Tecim, V., 2009. The use of GIS and multi-criteria decision-making as a decision tool in forestry. ORInsight 22 (2), 105–123.

Tinsley, C.H., Robin, L.D., Matthew, A.C., 2012. How near-miss events amplify or attenuate risky decision making. Manage. Sci.1–18 (published online ahead of print).

Tisdall, S., 2013. The Guardian, UK government must learn from Japan's catastrophe as it plans a new generation of plant, nuclear chief claims.

Titanic, 2000. The Sinking of the Titanic, 1912, Eye Witness to History. Available from: <www.eyewitnesstohistory.com> (accessed July 2010).

Toft, B., Rynolds, S., 1997. Learning from Disasters. Perpetuity Press, London.

Towill, D.R., 1999. "Management theory": is it of any practical use? Or how does a fad become a paradigm? Eng. Manage. J. 21 (2), 111–121.

Tsoukas, H., Chia, R., 2002. On organizational becoming: rethinking organizational change. Organ. Sci. 13 (5), 16.

Turner, B.A., 1978. Man-made disasters, Wykeham science series, 254p.

Turner, B.A., Pidgeon, N.F., 1997. Man-Made Disasters, second ed. Butterworth-Heinemann, London.

USDOI, Report regarding the causes of the April 20, 2010 Macondo well blowout. The Bureau of Ocean Energy Management, Regulation and Enforcement. United States Department of the Interior, USA, 2011.

Vaidogas, E.R., Juocevičius, V., 2008. Sustainable development and major industrial accidents: the beneficial role of risk-oriented structural engineering, technological and economic development of economy. Balt. J. Sustainability 14 (4), 612–627.

van de Ven, A.H., Poole, M.S., 1995. Explaining development and change in organizations. Acad. Manage. Rev. 20 (3), 510–541.

vanRee, C.C.D.F., Van, M.A., Heilemann, K., Morris, M.W., Royet, P., Zevenbergen, C., 2011. FloodProBE: technologies for improved safety of the built environment in relation to flood events. Environ. Sci. Policy 14, 874–883.

Vargas, L.G., 1990. An overview of the analytic hierarchy process and its applications. Eur. J. Oper. Res. 48, 2–8.

Vassou, V., Labib, A.W., Roberts, M., 2006. A decision model for junction improvement schemes. Proc. Inst. Civil Eng. Transport 159 (TR3), 127–134.

Vaughan, D., 1996. The Challenger Launch Decision. University of Chicago Press, Chicago, IL.

Vaughan, D., 2005. System effects: on slippery slopes, repeating negative patterns, and learning from mistake? In: Farjoun, M., Starbuck, W. (Eds.), Organization at the Limit: NASA and the Columbia Disaster. Blackwell, Oxford, UK.

Vaurio, J.K., 1984. Learning from nuclear accident experience. Risk Anal. 4 (2), 103–115.

Vinnem, J.E., 2013. Lessons from Macondo Accident, Chapter 5 165–177, Offshore Risk Assessment, vol. 1, Springer Series in Reliability Engineering.

Weick, K.E., 2003. Positive organizing and organizational tragedy. In: Cameron, K.S., Dutton, J.E., Quinn, R.E. (Eds.), Positive Organizational Scholarship: Foundation of a New Discipline. Berrett-Koehler, San Francisco, CA (Chapter 5).

Weightman, M., Learning from Fukushima, Physics World.com (online journal), 2012. Available from: <http://physicsworld.com/cws/article/print/2012/mar/06/lessons-from-fukushima> (accessed 12.12.13).

Weightman M., 2011. Japanese Earthquake and tsunami: implications for the UK Nuclear Industry—Interim Report, Office for Nuclear Regulation.

Weightman, M., Report: HM Chief Inspector of Nuclear Installations, Japanese earthquake and tsunami: implications for the nuclear industry. Interim Report, 2011. Available from: <http://www.hse.gov.uk/nuclear/fukushima/interim-report.pdf> (accessed 16.07.13).

Weir, D., 1987. The Bhopal Syndrome: Pesticides, Environment and Health. Sierra Club Books, San Francisco, CA.

Willmott, P., 1994. Total Productive Maintenance. The Western Way. Butterworth Heinemann Ltd., Oxford.

Winter, S.G., 1964. Economic "natural selection" and the theory of the firm. Yale Econ. Essays 4, 225–272.

Womack, J., Jones D.T., Roos, D., 1990. The Machine That Changed the World.

Wood, W., Lundgren, S., Ouellette, J.A., Busceme, S., Blackstone, T., 1994. Minority influence: a meta-analytic review of social influence processes. Psychol. Bull. 115, 323–345.

Woods, D.D., 2006. In: Hollnagel, E., Woods, D.D., Leveson, N. (Eds.), Essential Characteristics of Resilience. Ashgate, England.

Woods, D.D., Hollnagel, E., 2006. Prologue: resilience engineering concepts. In: Hollnagel, E., Woods, D.D., Leveson, N. (Eds.), Resilience Engineering Concepts and Precepts. Ashgate, England.

World Nuclear Association, Chernobyl Accident, revisited 2009. Available from: <http://www.world-nuclear.org> (accessed 17.07.11).

Xing, L., Amari, S.V., 2008. Fault tree analysis. In: Misra, K.B. (Ed.), Handbook of Per formability Engineering. Springer, London, pp. 595–620.

Xuereb, M., 2008. Cruise Liner Industry in Valletta: The Geography of a New Tourism Niche. University of Malta.

Zahedi, F., 1986. The analytic hierarchy process: a survey of the method and its applications. Interface 16 (4), 96–108.

Zainudeen, N.A., Labib, A.W., 2011. Practical application of the decision making grid (DMG). J. Qual. Main. Eng. 17 (2), 138–149.

Zollo, M., 2009. Superstitious learning with rare strategic decisions: theory and evidence from corporate acquisitions. Organ. Sci. 20 (5), 894–908.

Zollo, M., Winter, S.G., 2002. Deliberate learning and the evolution of dynamic capabilities. Organ. Sci. 13, 339–351.

OTHER RELATED LITERATURE

Ahmed, R., Koo, J., Jeong, Y., Heo, G., 2011. Design of safety–critical systems using the complementarities of success and failure domains with a case study. Reliab. Eng. Syst. Saf. 96 (1), 201–209.

Akgu, A.E., Byrne, J.C., Lynn, G.S., Keskin, H., 2007. Antecedents and consequences of unlearning in new product development teams. J. Organ. Change Manage. 20 (6), 794–812.

Alardhi, M., Hannam, R.G., Labib, A.W., 2007. Preventive maintenance scheduling for multi-cogeneration plants with production constraints. J. Qual. Main. Eng. 13 (3), 276–292.

Apostolakis, G., Lemon, D., 2005. A screening methodology for the identification and ranking of infrastructure vulnerabilities due to terrorism. Risk Anal. 25 (2), 361–376.

Aro, P., Carlsen, J.-L., Rice, A., Seminario, M., Wright, M., McClelland, S., et al., 1985. The Report of the ICFTU-ICEF Mission to Study the Causes and Effects of the Methyl Isocyanate Gas Leak at the Union Carbide Pesticide Plant in Bhopal. ICFTU-ICEF, Brussels/Geneva.

Bakolas, E., Saleh, J.H., 2011. Augmenting defence-in-depth with the concepts of observability and diagnosability from control theory and discrete event systems. Reliab. Eng. Syst. Saf. 96 (1), 184–193.

Banuelas, R., Antony, J., 2006. Application of stochastic analytic hierarchy process within a domestic appliance manufacturer. J. Oper. Res. Soc. 58 (1), 29–38.

Bertolini, M., Bevilacqua, M., 2006. A combined goal programming—AHP approach to maintenance selection problem. Reliab. Eng. Syst. Saf. 91 (7), 839–848.

Bogard, W., 1989. The Bhopal Tragedy: Language, Logic, and Politics in the Production of a Hazard. Westview Press, San Francisco, CA.

Boyland, L., 2011. *The New Media Journal*, The Fukushima Dai-ichi Nuclear Power Station Disaster. Available from: <http://newmediajournal.us/indx.php/item/822> (accessed 15.12.13).

Cagno, E., Caron, F., Mancini, M., Ruggeri, F., 2000. Using AHP in determining the prior distributions on gas pipeline failures in a robust Bayesian approach. Reliab. Eng. Syst. Saf. 67 (3), 275–284.

Carnero, C., 2006. An evaluation system of the setting up of predictive maintenance programmes. Reliab. Eng. Syst. Saf. 91 (8), 945–963.

Chen, L., Cai, J., 2003. Using vector projection method to evaluate maintainability of mechanical system in design review. Reliab. Eng. Syst. Saf. 81 (2), 147–154.

Cook, R.I., Woods, D.D., 2006. Distancing through differencing: an obstacle to organizational learning following accidents. In: Hollnagel, E., Woods, D.D., Leveson, N. (Eds.), Resilience Engineering Concepts and Precepts. Ashgate, England.

Corporation R, ReliaSoft, 2007. System Analysis Reference: Reliability, Availability and Optimization. Publishing, Tucson.

Cowlagi, R.V., Saleh, J.H., 2012. Coordinability and consistency in accident causation and prevention: formal system theoretic concepts for safety in multilevel systems. Risk Anal.

Cox, L., 2002. Risk Analysis Foundations, Models, and Methods. Kluwer Academic Publishers, Boston/Dordrecht/London.

Dekker, S., 2006. Resilience engineering: Chroniciling the emergence of confused consensus. In: Hollnagel, E., Woods, D.D., Leveson, N. (Eds.), Resilience Engineering Concepts and Precepts. Ashgate, England.

Distefano, S., Puliafito, A., 2007. Dynamic reliability block diagrams VS Dynamic fault trees. In: Reliability and Maintainability Symposium, 2007. RAMS '07. Annual, Orlando, FL, pp. 71–76.

Ekaette, E., Lee, R.C., Cooke, D.L., Iftody, S., Craighead, P., 2007. Probabilistic fault tree analysis of a radiation treatment system. Risk Anal. 27 (6), 1395–1410.

Elliott, M., 2010. Selecting numerical scales for pairwise comparisons. Reliab. Eng. Syst. Saf. 95 (7), 750–763.

Ferdous, R., Khan, F., Sadiq, R., Amyotte, P., Veitch, B., 2009a. Handling data uncertainties in event tree analysis. Process Saf. Environ. Prot. 87 (5), 283–292.

Ferdous, R., Khan, F., Veitch, B., Amyotte, P., 2009b. Methodology for computer aided fuzzy fault tree analysis. Process Saf. Environ. Prot. 87 (4), 217–226.

Ferdous, R., Khan, F., Sadiq, R., Amyotte, P., Veitch, B., 2011. Fault and event tree analyses for process systems risk analysis: uncertainty handling formulations. Risk Anal. 31 (1), 86–107.

Ferdous, R., Khan, F., Sadiq, R., Amyotte, P., Veitch, B., 2012. Handling and updating uncertain information in bow-tie analysis. J. Loss Prev. Process Ind. 25 (1), 8−19.

Ferdous, R., Khan, F., Sadiq, R., Amyotte, P., Veitch, B., 2013. Analyzing system safety and risks under uncertainty using a bow-tie diagram: an innovative approach. Process Saf. Environ. Prot. 91 (1−2), 1−18.

Forman, E., 1990. Random indices for incomplete pairwise comparison matrices. Eur. J. Oper. Res. 48 (1), 153−155.

Gallucci, R., 2012. "What—me worry?" "Why so serious?": a personal view on the Fukushima nuclear reactor accidents. Risk Anal. 32 (9), 1444−1450.

Gupta, J., 2002. The Bhopal gas tragedy: could it have happened in a developed country? J. Loss Prev. Process Ind. 15 (1), 1−4.

Ha, J., Seong, P., 2004. A method for risk-informed safety significance categorization using the analytic hierarchy process and bayesian belief networks. Reliab. Eng. Syst. Saf. 83 (1), 1−15.

Ha, J., Seong, P., 2009. A human−machine interface evaluation method: a difficulty evaluation method in information searching (DEMIS). Reliab. Eng. Syst. Saf. 94 (10), 1557−1567.

Ho, W., Lee, C., Ho, G., 2010. Multiple criteria optimization of contemporary logistics distribution network problems. ORInsight 23 (1), 27−43.

HSE Health and Safety Executive Report, The Tolerability of Risk From Nuclear Power Stations. Available from: <http://www.hse.gov.uk/nuclear/tolerability.pdf> (accessed 20.10.12).

Ishizaka, A., Labib, A.W., 2011. Review of the main developments of AHP. Expert Syst. Appl. 38 (11), 14336−14345.

Johnson, C.W., 2003. Failure in Safety-Critical Systems: A Handbook of Accident and Incident Reporting. University of Glasgow Press, Glasgow, Scotland, ISBN 0-85261-784-4.

Joshua, S., Garber, N., 1992. A causal analysis of large vehicle accidents through fault-tree analysis. Risk Anal. 12 (2), 173−188.

Kannan, V., 2010. Benchmarking the service quality of ocean container carriers using AHP. Benchmarking Int. J. 17 (5), 637−656.

Khakzad, N., Khan, F., Amyotte, P., 2011. Safety analysis in process facilities: comparison of fault tree and Bayesian network approaches. Reliab. Eng. Syst. Saf. 96 (8), 925−932.

Khakzad, N., Khan, F., Amyotte, P., 2012. Dynamic risk analysis using bow-tie approach. Reliab. Eng. Syst. Saf. 104 (0), 36−44.

Khakzad, N., Khan, F., Amyotte, P., 2013a. Dynamic safety analysis of process systems by mapping bow-tie into Bayesian network. Process Saf. Environ. Prot. 91 (1−2), 46−53.

Khakzad, N., Khan, F., Amyotte, P., 2013b. Risk-based design of process systems using discrete-time Bayesian networks. Reliab. Eng. Syst. Saf. 109 (0), 5−17.

Krasich, M., 2000. Use of fault tree analysis for evaluation of system-reliability improvements in design phase. In: Reliability and Maintainability Symposium, 2000 Proceeding Annual, Los Angeles, CA, pp. 1−7.

Labib, A.W., 2003. Towards an intelligent holonic maintenance system. J. Main. Asset Manage. 18 (4), 5−12.

Labib, A.W., Yuniarto, N., 2005. Intelligent real time control of disturbances in manufacturing systems. J. Manuf. Technol. Manage. 16 (8), 864–889.

Lees, F., 1996. Loss Prevention in the Process Industries. Butterworth-Heinermann, Oxford.

Li, H., Apostolakis, G., Gifun, J., VanSchalkwyk, W., Leite, S., Barber, D., 2009. Ranking the risks from multiple hazards in a small community. Risk Anal. 29 (3), 438–456.

Lindell, M., Perry, R., 1990. Effects of the chernobyl accident on public perceptions of nuclear plant accident risks. Risk Anal. 10 (3), 393–399.

Linkov, I., Satterstrom, F., Kiker, G., Seager, T., Bridges, T., Gardner, K., et al., 2006. Multicriteria decision analysis: a comprehensive decision approach for management of contaminated sediments. Risk Anal. 26 (1), 61–78.

Loewen, E.P., To understand Fukushima we must remember our past: the history of probabilistic risk assessment of severe reactor accidents. Conference Notes, Sociedad Nuclear Mexicana Conference, 8 August 2011.

Lopez, F., Di Bartolo, C., Piazza, T., Passannanti, A., Gerlach, J., Gridelli, B., et al., 2010. A quality risk management model approach for cell therapy manufacturing. Risk Anal. 30 (12), 1857–1871.

Mallor, F., García-Olaverri, C., Gómez-Elvira, S., Mateo-Collazas, P., 2008. Expert judgment-based risk assessment using statistical scenario analysis: a case study—running the bulls in Pamplona (Spain). Risk Anal. 28 (4), 1003–1019.

Marseguerra, M., Zio, E., Librizzi, M., 2007. Human reliability analysis by fuzzy "CREAM". Risk Anal. 27 (1), 137–154.

Martins, M., Maturana, M., 2010. Human error contribution in collision and grounding of oil tankers. Risk Anal. 30 (4), 674–698.

Michal, R., 2011. Conference Reviews Impacts of Fukushima Daiichi, World Nuclear Fuel Cycle Conference 2011, Nuclear News.

Millet, I., 1997. The effectiveness of alternative preference elicitation methods in the analytic hierarchy process. J. Multi-Criteria Decis. Anal. 6 (1), 41–51.

Misra, K.B., et al., 1990. Use of fuzzy set theory for level I in probabilistic assessment. Fuzzy Sets Syst. 37, 139–160.

Myddelton, D.R., 2007. They Meant Well: Government Project Disasters. IEA, Institute of Economic Affairs, London.

Nonaka, I., 1994. A dynamic theory of organisational knowledge. Organ. Sci. 5 (1), 14–37.

Nuclear News Staff, 2011. Fukushima Daiichi after the Earthquake and Tsunami, Nuclear News—Special Report.

Nuclear Power Technology Development Section, 2009. Boiling Water Reactor Simulator with Active Safety Systems, User Manual.

NUREG/CR-2300, January, 1983. PRA (Probabilistic Risk Assessment), Procedures Guide, US Nuclear Regulatory Commission.

Özgen, D., Önüt, S., Gülsün, B., Tuzkaya, R., Tuzkaya, G., 2008. A two-phase possibilistic linear programming methodology for multi-objective supplier evaluation and order allocation problems. Inf. Sci. 178 (2), 485–500.

Park, K., Lee, J., 2008. A new method for estimating human error probabilities: AHP–SLIM. Reliab. Eng. Syst. Saf. 93 (4), 578–587.

Pareek, K., 1999. The managemnet did not adhere to safety norms. Interview. Down to Earth, 56.

Paté-Cornell, E., Dillon, R., 2001. Probabilistic risk analysis for the NASA space shuttle: a brief history and current work. Reliab. Eng. Syst. Saf. 74 (3), 345−352.

Paulos, T., Apostolakis, G., 1998. A methodology to select a wire insulation for use in habitable spacecraft. Risk Anal. 18 (4), 471−484.

Pentland, B.T., 1999. Building process theory with narrative: from description to explanation. Acad. Manage. Rev. 24 (4), 711−724.

Perris, T., Labib, A.W., 2004. An intelligent system for prioritisation of organ transplant waiting list. J. Oper. Res. Soc. 55/2, 103−115.

Perrow, C., 1999. Normal Accidents: Living with High-Risk Technologies, second ed. Princeton University Press, Princeton, NJ.

Porac, J.F., Thomas, H., 1990. Taxonomic mental models in competitor definition. Acad. Manage. Rev. 15 (2), 224−240.

Rathnayaka, S., Khan, F., Amyotte, P., 2011a. SHIPP methodology: predictive accident modeling approach. Part I: Methodology and model description. Process Saf. Environ. Prot. 89 (3), 151−164.

Rathnayaka, S., Khan, F., Amyotte, P., 2011b. SHIPP methodology: predictive accident modeling approach. Part II. Validation with case study. Process Saf. Environ. Prot. 89 (2), 75−88.

Rathnayaka, S., Khan, F., Amyotte, P., 2012. Accident modeling approach for safety assessment in an LNG processing facility. J. Loss Prev. Process Ind. 25 (2), 414−423.

Roberts, K.H., Madsen, P.M., Desai, V.M., 2005. The space between in space transportation: a relational analysis of the failure of STS-107. In: Starbuck, W.H., Farjoun, M. (Eds.), Organization at the Limit: Lessons from the Columbia Disaster. Blackwell Publishing, Malden, MA, USA.

Saaty, T., 1987. Risk—its priority and probability: the analytic hierarchy process. Risk Anal. 7 (2), 159−172.

Saaty, T., 1994. Highlights and critical points in the theory and application of the analytic hierarchy process. Eur. J. Oper. Res. 74 (3), 426−447.

Saaty, T., Vargas, L., Dellmann, K., 2003. The allocation of intangible resources: the analytic hierarchy process and linear programming. Socioecon. Plann. Sci. 37 (3), 169−184.

Saaty, T., Peniwati, K., Shang, J., 2007. The analytic hierarchy process and human resource allocation: half the story. Math. Comput. Modell. 46 (7−8), 1041−1053.

Saaty, T.L., 1988. The Analytic Hierarchy Process. Pergamon Press, New York, NY.

Saaty, T.L., 1994. Fundamentals of Decision Making and Priority Theory with the AHP. RWS Publications, Pittsburgh, PA.

Saunders, M., Lewis, P., Thornhill, A., 2009. Research Methods for Business Students. Pearson, Harlow.

Seattle Times, 2013. U.S. questioned safety of reactors long ago. Available from: <http://seattletimes.nwsource.com/html/nationworld/2014529117_quakereactors18.htm> (accessed 12.12.13).

Shain, A., Ranjbar, M., Abedi, S., 2011. Critical discussion on the relationship between failure occurrence and severity using reliability functions. Manage. Sci. Eng. 5 (1), 26−36.

Shrivastava, P., 1992. Bhopal. Anatomy of a Crisis, second ed. Paul Chapman Publishing, London.

Simon, H.A., 1992. What is an "explanation" of behavior? Psychol. Sci. 3, 150−161.

Skogdalen, J., Vinnem, J., 2012. Quantitative risk analysis of oil and gas drilling, using deepwater horizon as case study. Reliab. Eng. Syst. Saf. 100 (0), 58−66.

Smith, D., 1990. Beyond contingency planning: towards a model of crisis management. Ind. Crisis Q. 4 (4), 263−275.

Steele, K., Carmel, Y., Cross, J., Wilcox, C., 2009. Uses and misuses of multicriteria decision analysis (MCDA) in environmental decision making. Risk Anal. 29 (1), 26−33.

Stone, D., Lynch, S., Pandullo, R., 1995. Flares', (online), Available from: <http://www.gasflare.org/pdf/Flare_Type.pdf> (accessed April 2009.).

Tahir, Z., Burhanuddin, M.A., Ahmad, A.R., Halawani, S.M., Arif, F., 2009. Improvement of decision making grid model for maintenance management in small and medium industries, Proceedings of the International Conference of Industrial and Information Systems (ICIIS), Sri Lanka, pp. 598−603.

Taghipour, S., Banjevic, D., Jardine, A., 2011. Prioritization of medical equipment for maintenance decisions. J. Oper. Res. Soc. 62 (9), 1666−1687.

Tavana, M., 2005. A priority assessment multi-criteria decision model for human spaceflight mission planning at NASA. J. Oper. Res. Soc. 57 (10), 1197−1215.

Thekdi, S., Lambert, J., 2011. Decision analysis and risk models for land development affecting infrastructure systems. Risk Anal. 32 (7), 1253−1269.

Ting, S.-C., Cho, D., 2008. An integrated approach for supplier selection and purchasing decisions. Supply Chain Manage: An Int. J. 13 (2), 116−127.

Todinov, M., 2006. Reliability analysis based on the losses from failures. Risk Anal. 26 (2), 311−335.

Tokyo Electric Power Company, 2011. Fukushima Nuclear Accident Analysis Report—Interim Report, TEPCO Web Publication.

Tokyo Electric Power Company, 2011. Reactor, Units 2 and 3, Fukushima Daiichi NPS Existence of the Impact by aged Deterioration Immediately after the Occurrence of Tohoku—ChihouTaiheiyou-Oki Earthquake, Interim Report, TEPCO Web Publication.

U.N. Scientific Committee on the Effects of Atomic Radiation, 2011. 36. Sources and Effects of Ionizing Radiation—Annex D Health effects due to radiation from the Chernobyl accident.

Varma, D., Mulay, S., 2009. Methyl isocyanate: the Bhopal gas. In: Gupta, R. (Ed.), Handbook of Toxicology of Chemical Warfare Agent. Elsevier, New York, NY.

Varma, R., Varma, D., 2005. The Bhopal Disaster of 1984. Bull. Sci. Technol. Soc. 25 (1), 37−45.

Vesely, W.E., Goldberg, F.F., Roberts, N.H., Haasl, D.F., 1981. Fault Tree Handbook (NUREG-0492). US Nuclear Regulatory Commission, Washington, DC.

Wachholz, T., 2012. The Eastland Disaster, first ed. Arcadia Publishings.

Weightman M., 2011. The Great East Japan Earthquake Expert Mission, IAEA International Fact Finding Expert Mission of the Fukushima Dai-ichi NPP Accident Following the Great East Japan Earthquake and Tsunami, Mission Report IAEA.

Xiao, W., Liu, Z., Jiang, M., Shi, Y., 1998. Multiobjective linear programming model on injection oilfield recovery system. Comput. Math. Appl. 36 (5), 127−135.

Yang, Z., Wang, J., Bonsall, S., Fang, Q., 2009. Use of fuzzy evidential reasoning in maritime security assessment. Risk Anal. 29 (1), 95−120.

Youngblood, R., 1998. Applying risk models to formulation of safety cases. Risk Anal. 18 (4), 433−444.

Yuniarto, N., Labib, A.W., 2005. Optimal control of an unreliable machine using fuzzy logic control: from design to implementation. Int. J. Prod. Res. 43 (21), 4509−4537.

Yuniarto, N., Labib, A.W., 2006. Fuzzy adaptive preventive maintenance in a manufacturing control system: a step towards self-maintenance. Int. J. Prod. Res. 44 (1), 159−180.

Zio, E., Baraldi, P., Popescu, I., 2008. A fuzzy decision tree for fault classification. Risk Anal. 28 (1), 49−67.

Glossary of Terms

AHP	Analytical hierarchy process
BWR	Boiling water reactor
CBM	Condition-based maintenance
DMG	Decision-making grid
DO	Design-out
FC	Favorable condition
FMEA	Failure mode effect analysis
FTA	Fault tee analysis
FTM	Fixed time maintenance
HRO	High reliability organization
IS	Investigative strategy
JIT	Just in time
MCDM	Multiple criteria decision making
MIC	Methyl iso cyanide
NAT	Normal accident theory
NPP	Nuclear power plants
PM	Preventive maintenance
RBD	Reliability block diagram
RCM	Reliability centered maintenance
RPN	Risk priority number
SU	Skill-level upgrade
SRB	Solid rocket boosters
TPM	Total productive maintenance

Index

F

Printed and bound by CPI Group (UK) Ltd, Croydon, CR0 4YY

03/10/2024

01040321-0002